Politics of the Wild

Canada and Endangered Species

Edited by

Karen Beazley and Robert Boardman

OXFORD

UNIVERSITY PRESS

OXFORD
UNIVERSITY PRESS

70 Wynford Drive, Don Mills, Ontario M3C 1J9
www.oupcan.com

Oxford University Press is a department of the University of Oxford.
It furthers the University's objective of excellence in research, scholarship,
and education by publishing worldwide in

Oxford New York

Athens Auckland Bangkok Bogotá Buenos Aires Cape Town Chennai Dar es Salaam
Delhi Florence Hong Kong Istanbul Karachi Kolkata Kuala Lumpur Madrid Melbourne
Mexico City Mumbai Nairobi Paris São Paulo Singapore Taipei Tokyo Toronto Warsaw

with associated companies in Berlin Ibadan

Oxford is a trade mark of Oxford University Press
in the UK and in certain other countries

Published in Canada
by Oxford University Press

Canadian Cataloguing in Publication Data
Main entry under title:
Politics of the wild : Canada and endangered species
Includes bibliographical references and index.
ISBN 0-19-541506X
1. Endangered species – Canada. 2. Endangered species – Government
policy – Canada. I. Beazley, Karen. II. Boardman, Robert.
QH77.C3P64 2001   333.95'42'0971   C00-933072-0

Cover design: Brett Miller

1 2 3 4 - 04 03 02 01
This book is printed on permanent (acid-free) paper ∞.
Printed in Canada

Contents

Preface

David T. Suzuki

At the end of World War II, my family was sent to southern Ontario where my parents worked as labourers on a fruit farm. In 1946, when I was 10, we spent a day at the Detroit Zoo and I vividly remember being overwhelmed by the variety of animals on earth. For years after, my daydreams were suffused with yearning to see the herds of animals on the Serengeti, to experience the Amazon with its air filled with the colour of butterflies and parrots, and to explore Australia's outback in search of a duck-billed platypus.

It's been my privilege to have visited those places of my boyhood dreams and more—Papua New Guinea, the Kalahari Desert, the Arctic, the Galapagos Islands, and Siberia. Whenever I visit a new area, I seek out elders so I can ask them what they remember of this place when they were children. And everywhere, the answer is the same chilling refrain: 'It used to be so different.' They speak of vast tracts of forests rich with animals, skies filled with birds, and waters teeming with fish. All over the planet, including its remotest parts, our elders are a living record of enormous ecological changes triggered by human activity. The scale and scope of these impacts have induced a massive loss of species and their habitats. The cause is called progress, a frenzied effort to exploit elements of the natural world far beyond their capacity to regenerate themselves.

Canadians pride themselves for living in a country of wild landscapes and bounteous life forms. I don't think it's an accident that Canada is where Greenpeace was formed or that seals, whales, and old-growth forests have inspired hundreds of thousands of Canadians to raise their voices to protect them.

But today, most of us live in cities. We inhabit a human-created environment of concrete, steel, and glass, where shopping malls and chain stores fulfil our every need in climate-controlled comfort. It's as if the seasons and the rest of life's diversity are no longer relevant to urban life. Recently, at a fund-raising event in Calgary, I was seated next to a wealthy businessman who was a fervent outdoorsman. During the evening, I described how grizzly bears had once been plains animals that lived on bison and ranged from Manitoba to Texas but now we think of them only as mountain animals because they've been extirpated everywhere else. 'I'm glad', he responded, 'because otherwise I'd really find it unpleasant to go hiking.'

It may seem incredible, but I am occasionally asked 'Who needs nature?' The question reflects the fact that the person asking it has forgotten the simple truth that no matter how sophisticated and removed from nature we seem to have become, we're still animals with an inescapable need for clean air, water, soil, and energy. And those fundamental biological needs are cleansed, renewed, and increased by the web of all life on earth that is now called biodiversity.

Today we have adopted a terrible illusion that we can manage wilderness and wildlife. Let's consider something much simpler than an ecosystem, say a candy factory. In order to manage it sustainably, we would need two fundamental bits of information: an inventory of everything in that factory and a blueprint indicating how all the components of that inventory are interconnected. It should be possible to do that with any human-created system, but we can't come close to that with even the simplest ecosystem. Estimates of the number of species on the planet range between 2 million and 100 million, with most biologists settling on 10–15 million. To date, biologists have identified about 1.5 million and identification simply means that someone working with a dead specimen has classified it as a new species and assigned it a name. It doesn't mean there is any fundamental information about that species—life cycle, habitat, food requirements, reproduction, interaction with other species, range, etc. The eminent biologist, Harvard's E.O. Wilson, suggests we have that kind of detailed biological information on a fraction of 1 per cent of all those species we have classified. With an inventory of perhaps 15 per cent of earth's biodiversity and a minuscule knowledge of their interconnections, how can anyone possibly suggest that we can manage any aspect of nature? The one component that we have a hope of managing is ourselves. We are the dominant predator on the planet and that depredation must be brought into balance with the regenerative capacity of the earth.

Extinction is natural; in fact, it is essential for life to evolve and change over time. Of all species that have ever existed, it's estimated that 99.9999 per cent are now extinct. What is unnatural about today's extinction crisis is the rate. Biologists consider the present eco-crisis as another massive wave of extinction to join five other periods known from the fossil record when significant numbers of species disappeared. The current annual extinction of 35,000 to 50,000 species will eventually level off and we know from the fossil record that the variety will increase again. But it will take 10–15 million years for the number of species to recover to the level before the crisis. Human beings are now the cause of the current extinction crisis and we have been around as a distinct species for perhaps 200,000 years. We've come a long way in a short time, but millions of years will be a long time for future generations to wait for Earth to recover.

Who needs nature? No species needs it more than we do. It's not just for the obvious fact that as animals we depend on nature's services for our very survival. Even if we are so clever that we will be able to create technologies to replace what we lose, we will have suffered an irreplaceable loss of spirit.

It grieves me to know that human beings deliberately exterminated bison and passenger pigeons, thereby denying all subsequent generations of humanity of experiencing the sheer mass and grandeur of those animals and their movements. In my lifetime, I've watched the great abundance of nature around Vancouver being drawn down so my grandchildren will never experience it. Instead, they are left to read or watch videos about the great oolichan runs, abundant abalone, and fabled salmon rivers.

So what inspires young people today? What replaces the dreams of wildebeest, jaguars, orchids, and orangutans? Is it surfing the Net, store-hopping in malls, playing the latest video games? Will future generations find the magic and mystery I experienced as a boy mucking about in swamps, gathering salamander eggs in ponds, or collecting insects in an empty field? I believe we desperately need nature to remind us of our real home, the web of living beings we share the planet with. We need a sense of wonder and awe at the magnificence and immensity of the cosmic forces impinging on our lives. We need to know that there are mysteries we will never solve, forces far beyond our ability to control or manage. Nature informs us of that all the time.

This important book shows us that the wild and wilderness have values that far surpass material necessity or economics. This book reveals to us the political and social challenges of protecting wilderness and outlines concrete steps that have to be taken. If wildness has the values I suggest, then this book should be crucial reading for all who care.

Abbreviations

BCAG	Biodiversity Convention Advisory Group
BCIC	Biodiversity Convention Interdepartmental Committee
CBD	Convention on Biological Diversity
CCAD	Canadian Conservation Areas Database
CCEA	Canadian Council on Ecological Areas
CEC	Commission for Environmental Co-operation
CEPA	Canadian Environmental Protection Act
CESC	Canadian Endangered Species Coalition
CESCC	Canadian Endangered Species Conservation Council
CESPA	Canadian Endangered Species Protection Act
CFS	Canadian Forestry Service
CITES	Convention on International Trade in Endangered Species of Wild Fauna and Flora
COHPS	Canadian Ocean Habitat Protection Society
COP	Conference of the Parties (CITES)
COSEWIC	Committee on the Status of Endangered Wildlife in Canada
CPAWS	Canadian Parks and Wilderness Society
CWS	Canadian Wildlife Service
DFO	Department of Fisheries and Oceans (Canada)
DPS	distinct population segment
EC	European Community
ESA	Endangered Species Act (US)
EU	European Union
GDP	gross domestic product
GIS	geographical information systems
GPI	genuine progress index
IFAW	International Fund for Animal Welfare
INC	Intergovernmental Negotiating Committee
ITTA	International Tropical Timber Agreement
IUCN	International Union for the Conservation of Nature and Natural Resources (World Conservation Union)

IWC   International Whaling Commission
MBS   Migratory Bird Sanctuary
MPA   Marine Protected Area
NAEC   National Agriculture Environmental Committee
NAFTA North American Free Trade Agreement
NGO   non-governmental organization
NIEO   new international economic order
NMCA   National Marine Conservation Area
NWA   National Wildlife Area
OBO   Operation Burrowing Owl
OECD   Organization for Economic Co-operation and Development
RCMP   Royal Canadian Mounted Police
RENEW Recovery of Nationally Endangered Wildlife
SAMPAA Science and Management of Protected Areas Association
SARA   Species at Risk Act
SARWG Species at Risk Working Group
SSSI   Site of Special Scientific Interest (UK)
TRAFFIC Trade Records Analysis of Fauna and Flora in Commerce
UNEP   United Nations Environment Program
UNESCO United Nations Educational, Scientific and Cultural Organization
USFWS   United States Fish and Wildlife Service
USGS   United States Geological Service
VEC   valued ecosystem component
WCED   World Commission on Environment and Development
WHC   Wildlife Habitat Canada
WMCC   Wildlife Ministers Council of Canada
WWF   World Wildlife Fund (also known as World Wide Fund for Nature)

Introduction

KAREN BEAZLEY AND ROBERT BOARDMAN

Never again will we delight in the singular beauty of sea mink, great auk, passenger pigeon, and blue walleye, for these species, which once roamed Canadian soil, sea, and sky, no longer exist here or anywhere in the world. In relatively recent times, geologically speaking, 12 species that used to exist in Canada have become extinct. Another 15 species no longer exist in the wild in Canada but occur elsewhere. In Canada there are currently 353 species and populations listed as extinct, extirpated, endangered, threatened, or vulnerable (COSEWIC, 2000). The present rate and magnitude of species extinction is unprecedented in geological history since the loss of the dinosaurs and other species 65 million years ago. The current global loss of species has reached a dramatic level, that of an extinction spasm, and its primary cause, unlike earlier events, is human activity. Extinction and the endangerment of species are a major component of the biodiversity crisis, a systematic and perhaps cataclysmic reduction in the diversity of life on earth. Between 10 and 40 per cent of all species on earth are already on the verge of extinction (Pimm in Suzuki and Dressel, 1999: 26). Further, we are threatened with the loss of 30–50 per cent of earth's biodiversity, fully one-third to one-half of all species that currently exist (Loucks in Suzuki and Dressel, 1999: 28; Myers, May, Jablonski, Reid, and others in Quammen, 1998). Yet all species, it may be argued, from the most humble fungi to the largest of carnivores, are essential components of communities, ecosystems, evolution, and our life-support system.

David Quammen (1998) summarizes the predominant causes of the present extinction episode as continuing landscape conversion resulting in habitat loss and fragmentation, the growth curve of the human population, and further ecological dislocations caused by invasive species. The wild places that remain intact, even those that are protected, are severely jeopardized by insularization, the needs of the world's poor for food, invasive species, and climate change. Quammen describes the partitioning and unifying of the world's landscapes, by habitat fragmentation and the global transport of weedy invasive species, as two converse trends with one doubled result: the accelerating loss of biological diversity. He paints a picture of a future in which nature 'will look very different'—'virtually everything will live virtually everywhere, though the list of species that constitute "everything" will be small', and lions, tigers, and bears will exist only in zoos. He characterizes this world as a 'planet of weeds', in which some privileged humans are likely to survive by relying on expensive stop-gap technologies in an otherwise biologically, economically, and spiritually impoverished world rife with new difficulties.

By Quammen's and others' theses, the situation seems hopeless. Even if we were to change significantly our attitudes, policies, and practices in the very near future, a substantial proportion of earth's biodiversity would still be lost. The biodiversity crisis is well under way, the causes are complex, and the results are predicted to be devastating. This is the situation we find ourselves in and this is the reason for this book: to raise awareness of the magnitude of the crisis and of the political, philosophical, scientific, and other challenges we face in responding to it. The hope lies in the struggle.

The endangerment of species is fundamentally a result of myriad social and cultural assumptions, political and economic systems, lack of understanding of the interrelationships involved, the complexity of the issues, and conflicting values among various interests. Further, the impacts are a result of incremental and cumulative factors, occurring over time and in ways in which the relationships between cause and effect are not always readily or immediately apparent. Declines of populations of species often occur over long time frames, across large spaces, and otherwise outside of the normal context of general human observation.

Thus, the biodiversity crisis is typical of what Bryan Norton has described as a 'third generation problem' (1995). Third-generation problems such as cataclysmic loss of biodiversity share complex features. Generally, they are large-scale, cumulative, and systematic phenomena with catastrophic effects, resulting from myriad small but incremental decisions, that harm large numbers of individuals over broad geographical areas, mainly in future generations. As a result, serious ethical issues arise regarding our responsibilities to future generations of humans and to other species. These are the ones who will bear a disproportionate burden of the costs associated with an impoverished planet without reaping any of the benefits we are currently enjoying. Dilemmas such as these are very difficult to address and are unprecedented in scale and complexity for human response and management. Efforts in Canada and elsewhere to respond in effective ways to protect endangered species specifically, and biological diversity generally, serve to illustrate the significant ethical, scientific, and political challenges involved. The Convention on International Trade in Endangered Species of Wild Fauna and Flora (CITES) and the Convention on Biological Diversity are examples of international agreements in response to these global issues.

Habitat conservation is intimately related to the issue of endangered species. As noted earlier, landscape conversion and fragmentation are primary causes of extinction. In Canada, as elsewhere, the finite land base and the surrounding sea are increasingly subject to competing forces. Land is continually and incrementally being converted from natural ecosystems and habitats to areas dominated by human activities such as agriculture, urbanization, and various forms of industrialization. Forests are being harvested at unprecedented and often unsustainable rates, and increasingly remote areas are being subject to mineral and energy exploration and development. Rising human populations and expectations exacerbate this situation. These multiple pressures on habitats are having direct and significant effects on species.

One response to the call to protect species and habitat was the World Wildlife Fund Canada's Endangered Spaces Campaign. The initiative received widespread public support and consequently the federal government made a commitment to set aside representative samples of each natural region in Canada in a national park by the year 2000. Many provinces also made this commitment, setting aside representative samples of typical landscapes in protected-area systems.

These initiatives represent a relatively new emphasis on protection of whole ecosystems and a shifting away from a focus on individual species in conservation initiatives. This is an understandable and effective strategy; habitats must be conserved to protect species. Species cannot survive without their habitats. Further, ecosystem-based approaches capture several species at a time, protecting whole communities rather than individual species, thus they can be more efficient or effective and do not require a thorough understanding of the specific habitat requirements of viable populations of every individual species. Some species, however, will not necessarily be adequately protected within areas set aside to represent typical landscapes or ecosystems. These include species with specific habitat or large area requirements or other distinguishing characteristics that make them vulnerable to extinction. Further, the ecosystem is or should be defined by the needs of the species that comprise the living component of that ecosystem. Ultimately, an approach informed by both species and ecosystem considerations is essential to any effective conservation strategy.

Although it is important to be concerned about all species, certain species are more at risk than others. All species at risk may be referred to in a general way as endangered species. Within this book the term 'endangered species' usually refers, in this general way, to all species at risk in Canada. However, Canada also has various specific categories of species at risk that characterize or define the degree of endangerment or risk of extinction. Species that no longer exist are listed as 'extinct'. 'Extirpated' species no longer exist in the wild in Canada, but they occur elsewhere. 'Endangered' species are facing imminent extinction or extirpation. 'Threatened' species are likely to become endangered in Canada if limiting factors are not reversed. 'Vulnerable' species are of special concern because of characteristics that make them particularly sensitive to human activities or natural events (Canadian Wildlife Service, 1997).

At both national and provincial levels, Canada has various pieces of legislation that deal with wildlife and endangered species of wild flora and fauna. Most provinces have some sort of wildlife legislation and the Canadian Wildlife Service is concerned with migratory waterfowl and other species that cross international boundaries. In Canada the federal government has relatively little jurisdiction over endangered species compared with the provinces. Management of wildlife, like other natural resources, has historically been left largely to the jurisdiction of the provinces. However, at the federal level, Canada has played a lead role in and is signatory to various international initiatives concerned with endangered species, such as CITES and the Convention on Biological Diversity. As such, Canada is under some obligation to establish federal endangered species and other legislation. Further, a large percentage of Canadians want strong endangered species legislation at the national level and believe that the federal government should take this lead role.

The international and national political contexts interact in complex ways. Problems in the harmonization of federal and provincial legislation in Canada, international pressures, the voices of First Nations and provincial and territorial governments, and the reality that species cross provincial and national boundaries all intersect to create a plethora of jurisdictional issues related to endangered species. Further, there is the interplay among non-governmental organizations (NGOs), industrial sectors, private landowners, labour groups, and other stakeholders, all with various perspectives on the issue, some of which conflict, but others of which are common among some of the groups. This has produced a situation in which consultation and compromise have resulted in a slow advance towards national endangered species legislation in Canada. Bill C-65 was introduced in 1996 but died on the order paper in 1997, at least in part because of a national election but also because the bill was not supported by many of the stakeholders. Bill C-33, the Species at Risk Act (SARA), was subsequently introduced in April 2000, but it, too, had its critics, and also died on the order paper with an election call in October of that same year.

Thus, this book is written within the context of a moving target and an unfinished political process. The issues, however, remain the same. Neither of the previously proposed acts was acceptable to all of the various players, nor were they sufficiently strong to serve the goal of protecting most endangered species. It is unlikely that the next proposal, assuming that there is one, will be significantly different in these respects, and thus the conflict will go on for some time. There is, as yet, no decisive resolution of the ongoing conflict over what should be done in Canada to protect endangered species.

Politics of the Wild represents the historical background and the context at this point in time. Views of the issue are presented from political, legal, ecological, philosophical, historical, geographical, and other perspectives. The book as a whole provides an integrated mix of themes related to the politics of endangered species and reflects the varied areas of expertise of the authors. It is essentially an interdisciplinary discourse on the politics of the wild in Canada as we move into a new millennium. As such, it is an important documentation of an exciting and evolutionary period in philosophical, ecological, and political history. It represents a melding of ethical, scientific, and political views in ways that anticipate the historical, contextual, strategic, and interdisciplinary approaches necessary to address the complex challenges of the future. It introduces a future that is currently facing us and the species with which we share the land and seas.

The book is organized into two main parts. The first part of the book looks at issues from the perspectives of philosophy and science, whereas the second part takes us from science to policy. The chapters in the first part, 'From Philosophy to Science', discuss the threats to Canada's wild species of fauna and flora and the principles governing sound practices of habitat protection and ecosystem management. In Chapter 1, Karen Beazley frames the biodiversity crisis in terms of endangered species and examines the rationale for conserving species. While it is important to protect habitat, there are also important philosophical and ecological reasons why we should, at the same time, focus on species. She identifies both instrumental and intrinsic arguments for protection of species at risk. In the first, species are regarded as valuable because they contribute to human needs and survival; in the second, the ethical case for protection is that species have inherent worth. Although the intrinsic value argument may ultimately be considered more philosophically sound, a case for enlightened self-interest is developed for the meantime. The final part of the chapter examines a focal-species approach as a broader alternative to biodiversity conservation, with the aim of preventing endangerment.

In Chapter 2, Bill Freedman, Lindsay Rodger, Peter Ewins, and David Green review the current status of Canada's species at risk and analyse the nature of the threats to them. Most risk, they point out, is caused by anthropogenic, or human induced, stress. Factors include the physical destruction of critical habitats, the effects of chemicals, and pressures from non-native species. Intense anthropogenic stressors are particularly important in the more densely populated southern regions of Canada. The authors discuss the data-gathering and recovery plans of, respectively, COSEWIC and RENEW. Among other points, they highlight the need for improved data (especially for species outside the traditional focus on vertebrates), a co-ordinated system of effective legislation, and greater environmental literacy on the part of Canadians.

David Gauthier and Ed Wiken describe in Chapter 3 the problem of endangered species as a 'negative legacy' for Canada. They argue that this is an indication of more fundamental problems of the health of habitats and ecosystems. We need sound ecological frameworks for conservation management and regionally based assessments to build the knowledge of how humans and other species can coexist. It would be an ineffective conservation approach to design management programs solely on the basis of individual species or even of communities. The specific requirements they identify are multiple-level partnerships among governments, landowners, and other groups; conservation strategies based on ecological principles; the integration of social, economic, cultural, and other factors into decision-making; and frameworks shaped by the principles of sustainable development and ecosystem management.

Chapters 4 and 5 discuss the problems of endangered species in relation to terrestrial and marine protected areas. Philip Dearden, in Chapter 4, observes that habitat destruction is the main cause of endangerment. The National Parks system plays a crucial role in the protection of rare and endangered species, for example, through the collection of data and the reintroduction of species. Dearden also discusses the contributions of provincial and other protected areas and of voluntary organizations, and the range and intensity of threats that continue to face wildlife. Using examples of programs such as those on the piping plover and burrowing owl, he argues that we need more protected areas, better links among networks of reserves, greater knowledge of species, and an ecosystem-based approach to management.

Martin Willison, in Chapter 5, argues that problems of marine species have traditionally been neglected because of the myth that marine systems are resilient and that human activities have minimal impacts on them. He discusses the record of marine protected areas in Canada, particularly in light of the provisions of the 1997 Oceans Act. As he shows through reference to several key examples, the protection of marine species is still inadequate. We have little data on many species. He concludes by discussing the prospects for the creation of an integrated network of marine protected areas.

The second part of the book, 'From Science to Policy', turns to the social, legal, and political contexts in which Canadians respond to the problems of species at risk. As Stephen Bocking explains in Chapter 6, these responses have a history dating back several decades. Before 1960, he writes, attention to wildlife meant primarily a celebration of its abundance. Changing attitudes and new programs followed the increasing appreciation of the threats to wildlife species from chemicals, loss of wetlands, and other developments. The provinces became more active players. COSEWIC was established in 1977, and RENEW in 1988. Species became a more political issue as the Canadian Wildlife Service and other organizations were increasingly drawn into programs of habitat protection. In particular, scientists and naturalists influenced events, he argues, as did growing awareness on the part of Canadians of the non-economic value of wildlife.

The federal government introduced a proposed endangered species law in 1996, which failed in 1997. The fate of Bill C-65, the Canadian Endangered Species Protection Act (CESPA), is analysed by William Amos, Kathryn Harrison, and George Hoberg in Chapter 7. They employ a 'policy regime framework' to explain policy outcomes as a function of three interacting components—actors, institutions, and ideas. They argue that Ottawa was unable to put together the required 'minimum winning coalition' that would have ensured passage of the bill. Key actors were opposed: environmentalists wanted more protection for species and habitats, private landowners and industry were critical of excessive regulation, and the provinces, territories, and First Nations objected to the powers the bill would have given to the federal government. Their analysis focuses on the impact of public opinion; the influence of ideas, including pressures from the scientific community; the role of key players, particularly the environmental community and industry and private landowners; and the significance of federalism and the interplay among Canada's governmental institutions. The chapter concludes with an overview of the main principles and provisions of SARA.

In Chapter 8, Robert Boardman looks at the issues of Canadian species at risk in the comparative context of other Western countries, particularly Australia, the United States, and the member states of the European Union. There are important similarities as well as differences in the patterns of threats to wildlife and the approaches of governments, voluntary organizations, and others to species protection and habitat management. As in Canada, NGOs have become crucial to recovery programs in many other countries; in several of these, too, the politics of endangered species is complicated by multiple jurisdictions. The historical absence of endangered species legislation at the national level, however, makes Canada's record among Western states anomalous. Boardman also discusses the increasingly crucial issue of the protection of species at risk on private land and the alternative strategies devised to manage this risk.

Much of what Canada does in relation to endangered species, and environmental policy more generally, has important connections with international developments. These are the focus of Chapter 9, by Philippe Le Prestre and Peter Stoett. Canada's international leadership role has traditionally been acknowledged by other countries. Further, our responsibilities to endangered species, they point out, extend beyond Canada's borders. The chapter focuses on the continuing importance of two key agreements: the Convention on International Trade in Endangered Species (CITES) of 1973, and the Convention on Biological Diversity (CBD) signed in 1992. Each is characterized by internal conflicts, for example, in CITES on the issue of harvesting species; similarly, the character of the CBD has reflected the concerns of the different constituencies involved in its creation. Both conventions remain significant influences on the way Canada's governments approach issues of species and habitats.

Finally, in Chapter 10, Robert Boardman, Amelia Clarke, and Karen Beazley discuss the factors of continuity and change that will likely mark approaches to issues of species at risk in the future. They review the developments surrounding the strengthening of the National Accord on Protection of Species at Risk and the consultations that led to the federal government's proposed Species at Risk Act in 2000. They argue that the factors that shaped endangered species debates and politics in the 1990s will continue to influence developments. These include the viewpoints of a wide range of players—scientists, environmental groups, labour, landowners, forestry companies, and the multiple governments and jurisdictions of Canada's political system. Effective national protection of wildlife species, they argue, and minimizing the threats that continue to endanger these are dependent on the development of sustained practices of co-operation among interested parties, along with a rethinking of individual and societal values regarding endangered and other species.

These chapters offer an overview of the context and complexity that pervade the politics of endangered species in Canada. These are important issues for Canadians. As a nation, we pride ourselves on our heritage of carving a nation out of the wilderness. Canada is caretaker of both a national and global legacy of wild flora and fauna, some of which exist nowhere else on earth. What remains of this rich natural heritage for future generations of both humans and non-humans will be a direct result of what we do today. Ultimately, it is no less a question than that of the future course of evolution. Today is a critical time in the politics of species at risk in Canada, the *Politics of the Wild.*

REFERENCES

Canadian Wildlife Service. 1997. *Endangered Species in Canada.* Ottawa: Environment Canada.

Committee on the Status of Endangered Wildlife in Canada (COSEWIC). 2000. *Canadian Species at Risk.* Ottawa.

Quammen, D. 1998. 'Planet of Weeds', *Harper's Magazine* 297, 1781: 57–69.

Norton, B. 1995. 'Environmental Problems and Future Generations', in J. Sterba, ed., *Earth Ethics.* Englewood Cliffs, NJ: Prentice-Hall, 129–37.

Suzuki, D., and H. Dressel. 1999. *From Naked Ape to Superspecies.* Toronto: Stoddart.

PART ONE

FROM PHILOSOPHY TO SCIENCE

Why Should We Protect Endangered Species? Philosophical and Ecological Rationale

KAREN BEAZLEY

INTRODUCTION

The human ideal of progress and the activities carried out in its honour reflect little regard for species and ecosystems. This has resulted in the current biodiversity crisis with which we are faced. Yet, despite centuries of denials and rationalizations, humans remain as dependent on and interconnected with the earth and its myriad species and processes as ever. We always will. Environmental philosophers, conservation biologists, and others hope that by focusing attention on the value of plant and animal species, as well as on the risk of extinctions, more ecologically responsible decisions will be made. Thus, humans and the other life forms with which we share this finite planet might continue to exist and evolve over time in mutually beneficial ways.

Framing the biodiversity crisis in terms of endangered species provides both benefits and special challenges. It is one way of focusing biodiversity management objectives on vulnerable species or species that warrant special management consideration. These species are in crisis situations and require immediate policy responses. It is also an understandable approach. Most people can identify more easily with individual creatures and species than they can with more abstract notions such as ecosystems.

Yet, there are implications of defining the biodiversity crisis in terms of endangered species. First of all, it is always a form of crisis management, reactive and requiring heroic measures, to bring species back from the brink of extinction. Such heroic measures require significant resources to implement, much more so than preventive measures, and draw scarce conservation resources away from prevention. Re-establishing a viable population of an endangered species requires much greater effort, if it is possible at all, than maintaining a healthy population in the first place.

Second, species become endangered for a variety of reasons, although the most common ones in North America are habitat loss and fragmentation, perhaps soon to be overwhelmed by dislocations caused by invasive species. It is a complex task to respond in a crisis situation to the multiplicity of factors at work. It is also difficult to redress the loss of habitat retrospectively and in a timely manner, more often than not requiring restoration, regeneration, or reclamation rather than the more simple prospect of habitat protection.

Further, decision-making in regard to endangered species is often made within the context of specific cases and posited in terms of so-called lifeboat scenarios. For example, decision-makers are asked to choose between potentially high revenue-generating or employment-generating developments or private property development rights and a few individuals of seemingly inconsequential endangered species such as a type of butterfly. 'Butterfly lovers' are often characterized as unreasonable, and decisions in favour of the butterfly are seen as unconstitutional.

Even if decisions involving endangered species are made in ways that are favourable to the species, the status quo remains for all other species. Species that are not listed as endangered remain subject to further persecution, exploitation, and habitat loss and degradation. An endangered species response to the biodiversity crisis alone does very little, if anything, to protect unlisted species, thus posing a risk of future endangerment to many more species. Rather, biodiversity policy could focus on identifying and protecting species that are potentially vulnerable to future endangerment. For example, species that experience population declines below the normal range of variation, habitat loss or degradation, or unsustainable levels of harvesting warrant special management attention in order to prevent endangerment.

The practice of focusing exclusively on endangered species or at the species level in general, rather than at the community or ecosystem level, may also be criticized. Focusing on higher levels of organization like ecosystems or landscapes serves to protect habitat and thus helps to prevent endangerment of species in the first place. Further, focusing on protection of representative and unique ecosystems serves to protect many species without the necessity of understanding the habitat requirements of every individual species.

While protecting habitat, ecological processes, and enduring features is important, protecting species, it may be argued, remains the *raison d'être* for these conservation objectives. Species are what comprise the living components of communities, ecosystems, and habitats and what we protect should be defined in terms of their needs. Aldo Leopold (1949), the father of holistic or ecosystem-wide ethics, recognized that to maintain the integrity, health, and beauty of the community requires that we keep all of the pieces. Individual populations of species are what will move, adapt, or perish in response to environmental change. Therefore, it is necessary to understand and protect species and their habitat requirements. It is also important to be able to identify the species upon which we should focus our attentions because it is not possible to focus on them all, nor is this necessary. Endangered species require immediate attention if we are to maintain their presence on earth.

Beyond the question of *how* to go about addressing the biodiversity crisis in terms of protecting endangered and other species is the question of *why* we should do so. Species constitute the living component of our life-support system. But do they have value beyond this, in and of themselves, regardless of their critical utility to us? This question is central to the politics of endangered species in Canada. How Canadians answer it influences the value context in which political decisions are made. An understanding of the philosophical and ecological reasons for valuing species in general, and endangered species in particular, is essential to making informed decisions affecting the future of species, both human and others, in Canada.

ENDANGERED SPECIES AS PART OF OUR ETHICAL SPHERE

The question of protecting endangered species is as much a philosophical or ethical issue as it is an ecological or political one. Indeed, it is perhaps even more so, because it has to do with our underlying belief systems. Ultimately, however, both philosophy and ecology are about understanding human identity in relation to the rest of the world. What are our roles and responsibilities in relation to other humans and to the non-human world? To justify our behaviour on philosophical or ecological or any other grounds is exercising and reflecting an ethic, whether we explicitly recognize it or not.

Two main realms of ethical argument for valuing endangered species and justifying their protection integrate ecology and philosophy. They are different from each other in a very fundamental way: in whether we value endangered species instrumentally or intrinsically. While these are fundamentally different ways of valuing species, they are not mutually exclusive. One may hold both sets of values simultaneously.

The first argument for valuing endangered species is related to their *instrumental* worth. Endangered species are part of the interrelated web of life that constitutes our life-support system; endangered species contribute, along with every species, to the overall diversity of life or biological diversity. Human survival depends on the continued existence of this life-support system; thus, endangered species have instrumental value. Therefore, humans should protect all species out of enlightened self-interest.

The second realm of argument for valuing endangered species is their *intrinsic* value or inherent worth. Endangered species exist, as products and processes of evolution. Endangered species, along with other species, have value in and of themselves, regardless of their utility to humans or anything else; thus, endangered species have inherent worth or intrinsic value. Therefore, humans have a moral obligation not to end a species line or to interfere with the right of other species to exist.

This dichotomy resonates with Michael Soulé's (1997) observation from years of teaching and working in the field of conservation biology. He says there are two groups of people within the environmental or conservation/preservation movement: (1) those who want to save the world to save the people; and (2) those who want to save the world to save the world. This reflects the dichotomy of purpose between instrumental value and intrinsic value and between an anthropocentric (human-centred) view and biocentric (life-centred) or ecocentric (life-centred plus non-life-centred) views. It is important to recognize these two perspectives, to place oneself within the appropriate sphere and to understand that other protectors of endangered species may not be in the same sphere, although we are all interested in saving the world.

Instrumental value has to do with use or utility: something has value because of its use or utility to someone or something else. We value the live bacteria cultures that we find in yogurt because they are good for our health. We value trees because they filter carbon dioxide and provide lumber, paper, and jobs. We value forests because they are spiritually or aesthetically pleasing and provide recreational opportunities. We value the biosphere because it provides our life-support system. These are all examples of instrumental values of species and ecosystems.

Intrinsic value recognizes that something has value regardless of its use or utility to us or anything else. It has value simply because it exists, like we do. We generally recognize that human beings have intrinsic value, or inherent worth, regardless of their usefulness to us or to society. This is recognized in sanctity-of-life arguments such as those around abortion and euthanasia and in human rights declarations. Within a broader view of intrinsic value, we recognize that all products of evolution have value, including those things that are not apparently useful and perhaps are even dangerous to humans and other forms of life, such as wood ticks and viruses.

The anthropocentric and non-anthropocentric perspectives can, however, be reconciled (Sterba, 1995b). First, all species, including humans, have a right to meet their basic needs by aggressing against the basic needs of other species. In other words, all individuals and species have a right to self-preservation. Individuals of all species have this right. This is not a human-centred approach. Very few species, however, aggress against the needs of other species or those species they depend on for survival to the extent that they eliminate those species. Humans have the population base, technological power, and economic value system for such behaviour on a global scale. This behaviour has been described as analogous to that of the cancer cell. Second, all species, including humans, have a right to defend or protect themselves against aggression by other individuals and species. This is analogous to the idea of self-defence in human morality and law. Individuals of all species have this right, as well as the right to self-preservation.

However, no individuals or species have the right to aggress against the basic needs of others to fulfil their own luxury wants. Sterba refers to this as the Principle of Disproportionality. No single individual or species has the right to a disproportionate share of the earth's resources at the expense of the basic survival needs of others. This is an issue of equity, and it transcends the boundaries of the human community to include the ecological community.

In very few instances in Canada, if any, can the aggression against the basic needs of species, to the point of extirpation or endangerment, be justified on the basis of meeting basic human survival needs or human self-defence. In almost all cases, if not all, species are being extirpated or endangered as a result of the fulfilment of human luxury wants. This type of behaviour would not be considered morally acceptable within the realm of human-to-human relations, nor is it morally justifiable in human relations with the rest of the ecological community. This does not mean that humans have to return to basic subsistence lifestyles. Rather, we must reassess our ideas of progress and wealth to include those things that truly contribute to a healthy and fulfilling lifestyle. These types of values are more readily indicated by a genuine progress index (GPI) than by the traditional economic measure of gross domestic product (GDP).

It is time to extend our moral consideration beyond the narrow confines of the human community to include the whole ecological community of which we are just one part. Our community of life does not stop at the boundary of the human species. Our ethical behaviour must be expanded to take this wider life-community into account in an equitable way. This is not simply for our own sake, but also out of respect for and awareness of the intrinsic value or inherent worth of other life forms, for their own sake, because they exist as both products and processes of evolution.

WHY ENDANGERED SPECIES MATTER: THE CASE FOR INTRINSIC VALUE

Holmes Rolston presents a philosophical argument on duties to endangered species that first appeared in *BioScience* in 1985; it is often reprinted in texts on environmental ethics. Rolston contrasts the instrumental and intrinsic bases of values and ethics in regard to endangered species, suggesting that in Western moral philosophy endangered species are basically valued instrumentally from four different perspectives:

1. Species are rivets in spaceship earth (for example, Ehrlich and Ehrlich, 1981).
2. Species are beautiful and awe-inspiring, like works of art (for example, Hargrove, 1987).
3. Species are important as economic, recreational, medicinal, and other resources.
4. Species are created by and declared good by the Creator, therefore they are sacred.

Rolston argues that these instrumental arguments are ultimately inadequate and he puts forward an alternative argument based in intrinsic value.

Rivets in Spaceship Earth

The first traditional argument examined by Rolston is that species are rivets in spaceship earth. This reflects the argument that all species are part of the interrelated web of life that forms our life-support system. Ehrlich and Ehrlich (1981) use the spaceship analogy to show that humans are removing rivets (species) from our life-support system at our peril. We simply do not know how many rivets can be lost before the spaceship starts to fall apart and crashes.

This argument is sufficient to justify the protection of some species, especially those that play important roles. Some ecosystems may collapse if functionally important species, sometimes referred to as keystone species, are lost. This issue is complicated by the fact that we do not really know which species, beyond some obvious ones, are functionally important. For example, Jim Estes (1998) describes the importance of key species at every trophic level (top carnivore, meso-carnivore, herbivore, producer or plants, and decomposer). Keystone species exist at each of these levels, and loss of any of them from any level has major implications for other levels and the system as a whole. (Such ecological considerations are discussed more fully in a following section of this chapter.) Further, we do not know how many species in total, keystone or not, we can lose before the community or system collapses. In the face of uncertainty and our lack of knowledge, prudence or the precautionary principle would suggest that we keep all the species. This, again, is analogous to Aldo Leopold's (1949) observation that the key to intelligent tinkering is to keep all of the pieces.

Rolston points out, however, that if it is possible to prove that a species is not important from the point of view of keeping the spaceship or the life-support system functioning, there is no rationale from this instrumental perspective for protecting the species. He correctly concludes that this argument only gives us reason to keep enough rivets to keep the ship from crashing. There is no reason or moral justification to save other species. Most species on the endangered species list and probably many others, particularly rare species, could be lost from spaceship earth without threat of ecosystem collapse. It is unlikely that our life-support system would collapse with the loss of most of the 353 Canadian species at risk (COSEWIC, 2000: ii). Clearly, the rivets-in-spaceship-earth argument for protecting endangered species is not adequate for justifying the protection of many of the currently endangered species.

Awe-inspiring

The second traditional view of morality that Rolston addresses in relation to species preservation is that species should be protected because they are beautiful or awe-inspiring or a source of wonder. Ultimately, this is about human enjoyment rather than the inherent worth of the species per se. Rolston (1985: 720) refers to this type of rationale as a 'fallacy of misplaced wonder'. We are valuing the human experience of wonder or awe or beauty rather than the objects of wonder, the species themselves. This is like valuing our experience of a painting by Van Gogh without valuing the work of art itself. (Even this is an inadequate analogy since the painting was created wholly from and for the human experience. In contrast, species were created from and for reasons and processes other than human needs or experience.) Further, this would be a case of aesthetic value, which is another example of instrumental value. And, again, this argument is insufficient to justify the protection of endangered species that may not be considered particularly beautiful or awe-inspiring.

Resources

Rolston also acknowledges that justification can be made for species preservation on economic, medicinal, and other resource or instrumental grounds: we need them, or may need them in the future, for food, raw material, or livelihood. However, many endangered species are not valuable as resources and most consist of too few individuals to be valuable on any large scale at the moment. Further, the economic system at present encourages, rewards, and justifies the liquidation of species, rather than protection or sustainability. Again, these are instrumental values and ultimately depend on proving the direct resource or utility value of a species to humans, a challenging prospect at best in the case of many endangered species.

The argument for preserving species on the basis of resource value is not only criticized by environmental philosophers. Peter Huber, a lawyer and senior fellow of the Manhattan Institute for Policy Research, suggests that the association of biodiversity with economic value is a case of 'an important environmental objective getting all tangled up in arguments that don't really wash' (1992, in Verburg, 1995: 18). He says that 'the economic case for biodiversity is weak . . . . We should revere life on earth not because we expect it will profit us economically, not because it is very likely to cure cancer, but because life is a good that requires no further justification' (ibid.).

Covenant with God
One final rationale that Rolston examines and that has traditionally been used to justify the protection of endangered species is of a spiritual or religious nature. It may be formulated such that God, Allah, Jehovah, the Goddess, the Great Spirit, the Creator, or the Divine created the earth and all species and declared them as good, and thus they are worthy of protection. Humans have a covenant with the creator not to destroy the created. For those of the Judeo-Christian tradition, this covenant with God is through Noah when he built the Ark; rainbows are reminders or symbols of this covenant. Such beliefs cannot be proven or justified through rationality. They must be accepted as a matter of faith. Rolston suggests that the difficulty with this style of argument is that one must hold the theological belief; it conveys no responsibility or offers no powers of suasion to those who do not accept the belief system.

All of these arguments or sources of justification or forms of morality related to endangered species will work for some people and in relation to some species. However, these arguments will not hold for species that are not keystone or functionally important, beautiful or awe-inspiring, or important economic resources, or with people who do not accept particular religious teachings.

Intrinsic Value
Rolston concludes, logically, that the only fundamental way consistently and reliably to justify human responsibilities towards endangered species is on the grounds of intrinsic value. Endangered species have intrinsic value or inherent worth simply because they exist, as both the products and processes of evolution, regardless of their utility to humans or anything else. Just as we accept that humans have inherent worth or intrinsic value regardless of their utility to us, other species also have this worth. All species have evolved from a common origin, and all have inherent worth and have it equally. This is essentially the only ultimate value: one that values all, regardless of utility. No one species has the right to extinguish the others. All have an equal right to exist.

Rolston is not alone in suggesting that forms of life are most fundamentally valued intrinsically because of their inherent worth as centres of life, pursuing a good of their own, whether consciously or not. Many other scientists, philosophers, and lawyers, along with conservationists, hold the view that the foremost argument for preserving all species of life on earth is because they have a right to exist. David Ehrenfeld named this argument the 'Noah Principle'. The Noah Principle suggests that every species has the same right to exist, the same inherent or intrinsic value, regardless of whether the species is considered to be a 'higher' or 'lower' organism, or of direct or indirect utility or disutility to humans or other species. Each has a right to exist; humans have a moral responsibility to each and every one (Ehrenfeld, in Mann and Plummer, 1992).

FOCUS ON THE SPECIES LEVEL

Another area of philosophical inquiry related to endangered species that Rolston addresses is why it is important to focus on the species level rather than on the individual level. Some say that if we take care of the individuals, the species will take care of itself. However, there are times in protecting endangered species when choices must be made between the good of the endangered species and the good of individuals of other species. How can one justify favouring endangered species over the lives of individuals of species that are not endangered? For example, sometimes it is necessary to cull a deer herd to save an endangered plant. Animal rights activists and others will argue against this type of approach because it violates the rights of individual creatures.

A species is not just a convenient taxonomic classification. Taxonomic classification represents or recognizes that the species is a life form that evolved as both a product and a process of speciation; thus, the species is both the product and the process of evolution. Although individuals are the means through which this life form exists and evolves, the species level represents a life form distinct from other life forms. Individual creatures live and die continuously in the process of speciation; dying is what allows speciation to occur. Rarely, if ever, is dying in the interest of the individual, yet it is necessary for the survival of species and for speciation. In contrast, ending a species line ends the speciating process; when this is done prematurely or artificially, it is a negative force in the evolutionary process. Preserving a species as a life form entails specific duties beyond those to individual creatures. According to Rolston, ending a life form is a kind of super-killing and requires super-justification. Preserving the forms preserves the formative process. And, since species cannot exist outside of their environment, it is species in communities, in ecosystems, that we want to protect.

Extinction: Natural and Artificial

There is also the question of extinction. Species regularly go extinct and at points in evolution there have been periods of mass extinction. However, there has not been a mass extinction of the rate and magnitude of the one we are experiencing now on earth since the episode in which the dinosaurs became extinct some 65 million years ago. That episode, theoretically, was triggered when a huge meteor smashed into the earth creating a 10-mile-wide crater on the Yucatán Peninsula of present-day Mexico, rocking the earth and creating massive tidal waves and dust and particle clouds that changed the climate (Wilson, 1999). In contrast, the current extinction crisis is being caused by the activities of humans. Regardless, some people will argue that extinction is a natural process and that the current extinctions are merely part of the evolutionary process. Further, they will suggest that we should not interfere with this natural process and that we have no responsibilities to stop a species from becoming extinct.

Rolston makes an important distinction between what he calls natural extinction and artificial extinction. Natural extinction occurs when a species is no longer suited to its environment. We have no duties to stop this sort of extinction. However, what we are seeing now are not natural extinctions. Humans are degrading and destroying habitats and over-harvesting species in ways that end the options of life forms that remain perfectly suited for their environment. This is artificial extinction. It is occurring well above the background rate of about one species in any major group every million years, and is predicted ultimately to affect between one-third and one-half of all species presently on earth (Quammen, 1998).

Human Superiority and Place in Nature

An extension of this argument is the suggestion that humans are part of nature, so therefore changes that we make to the environment and our impacts on other species are also natural. Further, our ability to dominate the world for our apparent advantage is understood to be because of, and further proof of, our superiority. Implicit in this argument is the assumption that superiority gives one the right to dominate, subjugate, or oppress another. This is a false and immoral assumption if we believe, for example, that a person who wins a running race does not have the right to oppress the other runners. The greatest rulers have been those who viewed themselves as serving or caring for their kingdoms or subjects. Yes, humans are part of the natural system and what we do impacts that system. Further, humans do have superior abilities in certain realms, such as the ability to think ahead and perceive the results of our actions (along with the additional benefits that opposable thumbs have wrought). It is exactly this characteristic ability for forethought that allows us to behave in a responsible manner—specifically, to behave as if we are part of the natural system, which we are, as opposed to somehow separate from or above it.

As humans, we have the abilities both to change the world and to perceive the results of our actions. With this comes a responsibility to act in ways appropriate to assure the best future. It is no less natural for us to behave as if we are a part of the ecological community and the process of evolution than it is to behave as if we are not. Perhaps it is time for us to evolve our consciousness, attitudes, and behaviour to reflect the reality of evolutionary processes and the ecological community. It is indeed ironic that the self-declared superior species with the ability for consciousness and moral consideration regards the welfare of only one species, Homo sapiens, as an object of morality or duty. As Rolston says, 'There is something Newtonian, not yet Einsteinian, besides something morally naïve, about living in a reference frame where one species takes itself as absolute and values everything else relative to its utility'

(1985: 726). We have the choice of behaving in an ecologically responsible manner that reflects our knowledge of the world and our place in it, or, as Aldo Leopold notes, we can carry on with the mentality of the potato bug, destroying the crop and, thus, ourselves.

ENLIGHTENED SELF-INTEREST, FOR THE MEANTIME

Protecting species out of recognition of their intrinsic value is, ultimately, the most philosophically sound argument and represents the ethical high ground. Gaining widespread support for this position at a societal level, however, will take some time. In the interim, enlightened self-interest, based on a set of instrumental arguments, could go a long way towards garnering greater public support for protection of species. Instrumental arguments alone, though somewhat less satisfying, could serve to justify the continued existence of most species, especially when we consider the current level of uncertainty in the state of our knowledge.

Regardless of the acceptance of intrinsic value, all of the species that exist on earth constitute the living component of our life-support system. All species, including humans, require the ecological processes and basic necessities of life that this support system provides, such as clean air, water, and sustenance. Further, these species serve as resources for other basic needs and as sources of livelihood and aesthetic and recreational pleasure. This is not to say that we have a right to use them to satisfy our non-basic or luxury wants or to fuel excessive consumption and the accumulation of wealth.

The diversity of species comprises an interrelated web of life, which consists of critical relationships and processes that we do not come close to understanding fully. The structure, function, and processes of this web of life and non-life are incredibly complex. Any choices we make regarding changes to this system would be fraught with uncertainty as to the effects on the whole or other parts of the system. Accordingly, a precautionary approach would dictate that we act conservatively and err on the side of caution, at the very least protecting all species from extinction. Indeed, a more rational approach would suggest that we maintain species at populations within their normal range of variation.

Within this context of general uncertainty, we do know of certain species that perform critically important functions and are sometimes referred to as keystone species. At the very least, these species, typically important prey or top predators at each trophic level, should be maintained to ensure these critical functions continue. Some known keystone species are currently endangered and others are vulnerable to future endangerment simply because of characteristics, such as larger body size, habitat area requirements, and position at the top of the food chain. While the number of keystone species currently endangered may be relatively low, more will be endangered in the future. Thus, we are not just talking about a few species, which we could perhaps afford to lose, but rather a significant segment of the ecological system. Further, we do not know which species are of critical functional importance; perhaps they all are. Along with the important role of keystone species, the diversity of life that all species provide is critical to maintaining health, stability, resilience, and redundancy within the life-support system.

Beyond the value that species provide for our current survival, there is the issue of our duties to future generations of humans, who have an equal right to the diversity of life not only from an aesthetic standpoint but as resource options and part of their planetary life-support system. Because humans are an opportunistic and resourceful species, some privileged groups are likely to survive in a future of post-ecological collapse by developing stop-gap technologies available at a price, such as bottled air and bottled water, as David Quammen (1998) suggests. But even this will not be without real difficulties in a biologically impoverished and degraded world, leaving many underprivileged humans and other species to perish. And who among us would be convinced that we would be among the survivors, and at what cost to ourselves, humanity, and the world? Perhaps, as David Jablonski suggests, some of these difficulties 'will serve as incentive for major changes in the trajectory along which we pursue our aggregate self-interests' (as cited in Quammen, 1998).

Thus, if a precautionary approach is taken, instrumental value alone provides sufficient justification, on the basis of enlightened self-interest, to protect endangered and other species. Each species may provide present and future life-support functions, as well as resources and aesthetic enjoyment. Beyond providing justification for the protection of endangered and other species, these arguments recommend the maintenance of populations of species within their normal ranges of variation.

ECOLOGICAL CONSIDERATIONS FOR THE PROTECTION OF SPECIES

Endangered species exist within a context of overall biological diversity on earth that includes the biosphere, ecoregions, and ecosystems, as well as communities, populations, and genes. Biological diversity requires the functional interplay among all these levels and all are important to conserve. Current protection initiatives in Canada tend to focus on protecting representative samples of natural regions within parks or protected areas. The problem with this focus is that some species (or populations) will not be protected or 'captured' within protected areas selected to represent typical examples of natural landscapes or regions, unique areas, or areas of scenic beauty or with recreational potential. Further, most existing and currently proposed protected areas are too small, too few, and too isolated to protect viable populations of some species. This is because the habitat requirements of viable populations of these species, such as habitat size and quality, have not been taken into account. Protection parameters defined at the ecosystem and landscape levels make no sense outside the requirements of the populations of the species that comprise that landscape or ecosystem.

At the same time, protecting adequate habitat, ecological processes, and enduring features is a good way to protect most species. The main causes of species extinction and endangerment are:

- habitat loss, fragmentation, and degradation;
- toxic effects of pollutants;

- mortality directly caused by humans, such as over-exploitation or persecution (i.e., many people kill wolves, coyotes, snakes, spiders, etc. simply because they don't like them or fear them);
- introduction of exotic species that out-compete native species for habitat space and other resources such as food.

Protecting ecosystems serves to address each of these concerns at least to some degree.

Therefore, it is important to understand the habitat requirements of species we are trying to protect and to incorporate these into ecosystem protection initiatives. It is also important to be able to identify which species should receive our greatest attention because there is no way we can understand the needs of all species, nor is this necessary. Some species, in landscapes and seascapes increasingly dominated and fragmented by human activities, are more at risk or endangered than others.

It makes sense to prevent species from becoming endangered in the first place. It would be useful to be able to predict which species are most likely to become endangered in the future. This is possible by identifying the types of characteristics that make certain species more vulnerable to endangerment. Each species can then be examined in turn to identify those that fit the characteristics. Several species in any particular region may possess characteristics that make them more vulnerable to extinction than others. These species, as well as currently endangered species, warrant special management attention.

Species that are endangered or most likely to become endangered, and thus most likely to benefit from special management attention and conservation measures such as protected areas, are:

- k-strategists, which are generally habitat or niche specialists, long-lived and large-bodied with low rates of reproduction and dispersal;
- summit or top predators, such as species that feed at the top of the food chain and often, as a result, concentrate toxins (most summit predators are also k-strategists);
- long-distance migratory species;
- species that concentrate spatially, such as congregating waterfowl, caribou, and musk-oxen; and
- large-bodied species (because of their generally low reproductive rates). (Theberge, 1993)

These most potentially vulnerable species may also be categorized as types of focal species (Hunter, 1990; Noss, 1990; Beazley, 1998; Miller et al., 1998–9). Categorization of focal species is useful for considering the different types of conservation responses required to protect these species. Four main types of focal species are keystone (functionally important), umbrella, habitat quality indicator, and flagship species.

Keystone or functionally important species are those such as top predators, important prey, major vegetation influencers, and those that change landscapes or seascapes, such as wolves, snowshoe hares, beavers, and earthworms. These species have a disproportionate importance in the function of the community in that the loss of these species could cause severe impacts on many other species and lead to ecosystem collapse or transformation. Protecting these species helps ensure that functional interactions among species in a community are maintained. Keystone species exist at every trophic level and it is important that each be maintained (Estes, 1998). Ecosystems missing any part of this trophic structure are functioning in states of imbalance and are not sustainable. It is important to maintain a range of species in each trophic level of interactions to maintain ecosystem functions.

Umbrella species are those with large area requirements or home range sizes because of their daily or seasonal movements, such as grizzly bears, caribou, moose, and salmon. These species are sensitive to loss and fragmentation of habitat area. If enough habitat area is maintained to support viable populations of these species, the habitat area requirements of many other species will also be protected.

Habitat quality indicators are species with special habitat or wilderness requirements, such as lynx, American marten, river otters, and pileated woodpeckers. These species are sensitive to loss, conversion, or degradation of particular specialized or undisturbed habitats such as forest interior, old-growth forests, tall grass prairies, or wetlands.

Flagship species are those, such as the bald eagle, moose, giant panda, and whales, that garner public support or political will for species conservation measures.

These types of species constitute a suite of focal species that warrant special conservation management attention. Together, they form a multi-species umbrella for conservation of many species. If the habitat and other management needs of these species are met, we will have gone a long way towards protecting most, if not all, species from endangerment. It is important to recognize that focal species should include those from all taxonomic classes, including plants and other invertebrate species, and terrestrial, aquatic, and marine species.

When a systematic assessment was conducted of the indigenous mammal, reptile and amphibian, and freshwater fish species throughout Nova Scotia relative to these criteria, all of the endangered species in these classes were identified (Beazley, 1998). This confirmed the notion that these are the right sorts of criteria to consider, that these are the characteristics that lead to vulnerability. The value of this type of assessment is that it also identifies those species currently not considered to be endangered but that will likely comprise the endangered species of tomorrow unless preventive measures are taken. They are species at risk of future endangerment. Thus, they warrant special management attention to prevent their continued decline.

In North America and other temperate regions, endangered subspecies and populations are important to consider, along with endangered species (Daily and Ehrlich, 1997–8). For example, black bear is not considered to be nationally or globally endangered. However, its historical range has been greatly reduced across North America.

This represents a significant loss of biodiversity, and it illustrates the need to recognize the importance of the consideration of appropriate scales and levels of endangerment and protection. It is important to conserve genetic and population diversity at both the inter- and intra-species levels. Thus, it is important to be concerned with protecting species at regional, provincial, national, and global scales.

In general, however, putting the focus of biodiversity conservation on endangered species is more often than not a case of too little too late—an act of desperation. It is at most a stop-gap measure and does little to protect or maintain adequate habitat to bring species back from the brink of extinction or to prevent species from becoming at risk. Most existing and proposed legislation and other initiatives for protecting endangered species do very little to address the causes of extinction, the main ones being landscape conversion, over-exploitation, and invasive species. Although specific legislation is required to protect endangered species and to identify those that are vulnerable, additional widespread measures at a more fundamental and societal level are also needed to address the causes of extinction and to prevent endangerment. Importantly, this includes valuing species for their inherent worth as well as for their utility to humans.

POSTSCRIPT

Perhaps there is no need to convince the readers of this text that protecting endangered species is the ethically and ecologically correct thing to do; this may be an instance of preaching to the converted. However, the philosophical issues raised here, and previously addressed by Rolston and others, are some of the tough questions with which we, as humans, are sometimes confronted.

Efforts to protect endangered and other species are, in and of themselves, a positive force in the evolutionary process. Individuals who engage in these efforts could be considered to be the living instruments of the formative process of human evolution, primarily one of consciousness. Tom Regan (1995: 72) says that 'The fate of animals is in our hands. God grant we are equal to the task.' Holmes Rolston (1985: 725) says that preserving the process of speciation, nature's tendency to produce diversity, is 'about as near to ultimacy as humans can come in their relationship with the natural world'.

So, we work towards a means of protecting species, especially endangered species, because we are compelled to do so. It is important to hold onto the idea of why we do the work, rather than to focus on the outcome. It is being a warrior, because we can do no less. Out of love, or, as Immanuel Kant would put it, positive inclination: to act beautifully, rather than out of a sense of duty. That is the good news. It can be a source of strength and inspiration in the face of an overwhelming task, given that as much as 40 per cent of all species on earth are currently endangered, and one-third to one-half of all species seem doomed to extinction before this episode ends. In the context of social, economic, political, and other institutional structures that are not set up to deal with such holistic and complex issues, we must pursue this task and maintain our focus on the process itself.

The other good news is this: the earth will survive this extinction event; life will go on, with or without us—though it may take 10–25 million years to recover (Livingstone, 1995; Quammen, 1998; E.O. Wilson and Norman Myers, cited in Suzuki and Dressel, 1999).

REFERENCES

Beazley, Karen. 1998. 'A Focal-Species Approach to Biodiversity Management in Nova Scotia', Ph.D. dissertation, Dalhousie University.

Committee on the Status of Endangered Wildlife in Canada (COSEWIC). 2000. *Canadian Species at Risk*. Ottawa.

Daily, G., and P. Ehrlich. 1997–8. 'Population Extinction and the Biodiversity Crises', *Wild Earth* (Winter): 35–45.

Ehrlich, P., and A. Ehrlich. 1981. *Extinction*. New York: Random House.

Ehrenfeld, D. Cited in Mann and Plummer (1992).

Estes, J. 1998. Keynote address delivered to The Wildlands Project Grassroots Rendezvous, Science and the Conservation of Nature, 8–11 Oct., Estes Park, Colo.

Hargrove. E. 1987. 'Eugene Hargrove: Foundations of Wildlife Protection Attitudes', *Inquiry* (Oslo) 30, 1, 2: 18–25.

Huber, P. 1992. Excerpt from 'Biodiversity vs. Bioengineering?', in Verburg (1995: 17–18).

Leopold, A. 1949. *A Sand County Almanac*. New York: Oxford University Press.

Livingstone, J. 1995. 'The Natural History of a Point of View', *The Nature of Things*, CBC television.

Mann, C., and M. Plummer. 1992. Excerpt from 'The Butterfly Problem', in Verburg (1995: 26–38).

Miller, B., R. Reading, J. Strittholt, C. Carroll, R. Noss, M. Soulé, O. Sanchez, J. Terborgh, D. Brightsmith, T. Cheeseman, and D. Foreman. 1998–9. 'Using Focal Species in the Design of Nature Reserve Networks', *Wild Earth* 8, 4: 81–92.

Noss, R. 1990. 'Indicators for Monitoring Biodiversity: A Hierarchical Approach', *Conservation Biology* 4, 4: 355–64.

Quammen, D. 1998. 'Planet of Weeds', *Harper's Magazine* 297, 1781: 57–69.

Regan, T. 1995. 'The Case for Animal Rights', in Sterba (1995a: 64–72).

Rolston, H. 1985. 'Duties to Endangered Species', *BioScience* 35, 11: 718–26.

Soulé, M. 1997. Keynote address delivered to the Canadian Council on Ecological Areas Conference, Protected Areas and the Bottom Line, 15–17 Sept., Fredericton.

Sterba, J., ed. 1995a. *Earth Ethics*. Englewood Cliffs, NJ: Prentice-Hall.

——. 1995b. 'Reconciling Anthropocentric and Nonanthropocentric Environmental Ethics', in Sterba (1995a: 199–213).

Suzuki, D., and H. Dressel. 1999. *From Naked Ape to Superspecies*. Toronto: Stoddart.

Theberge, J. 1993. 'Ecology, Conservation and Protected Areas in Canada', in P. Dearden and R. Rollins, eds, *Parks and Protected Areas in Canada*. Toronto: Oxford University Press, 137–53.

Verburg, C.J., ed. 1995. *The Environmental Predicament*. Boston: Bedford Books of St Martin's Press.

Species at Risk in Canada

BILL FREEDMAN, LINDSAY RODGER, PETER EWINS,
AND DAVID M. GREEN

INTRODUCTION

The conservation status of indigenous species in Canada is designated by a group of experts from government, academic institutions, and conservation organizations, known as COSEWIC (Committee on the Status of Endangered Wildlife in Canada). As of May 2000, COSEWIC has assigned 'at-risk' status to 353 indigenous taxa, of which 12 are *extinct*, 15 *extirpated* from Canada, 102 *endangered*, 71 *threatened*, and 153 of *special concern*. Compared with their relative contribution to the species-level biodiversity of Canada, vertebrate animals are highly overrepresented in the COSEWIC list, while non-vascular plants, lichens, fungi, invertebrates, and microorganisms are greatly underrepresented. Once a taxon is listed by COSEWIC, a parallel government-led body known as RENEW (Recovery of Nationally Endangered Wildlife) has been active in the preparation of recovery plans. As of 2000, RENEW has approved 17 recovery plans, equivalent to 4.5 per cent of the 353 taxa listed as extirpated, endangered, threatened, or of special concern in Canada (another 24 plans are in various stages of preparation).

Anthropogenic stressors are the main cause of virtually all cases of conservation risk to indigenous species in Canada. The most frequent cause of damage (affecting more than 80 per cent of the listed taxa) is the physical destruction of habitat. In addition, invasive non-native species are damaging the habitat of many taxa, or they are affecting them by intensive predation or competition. Physiological stress caused by anthropogenic emissions of toxic chemicals is also an important stressor of many species at risk. Harvesting is an additional stress to populations of species of economic value. A disproportionate fraction of species at risk occurs in southern regions of Canada, where the number of indigenous species is high, habitat loss has been most extensive, and the intensity of other anthropogenic stressors is greatest (for example, pollution and harvesting). More effective actions are urgently needed to prevent and mitigate the damage being caused to species at risk in Canada. Especially crucial is the need for effective federal and provincial/territorial legislation that would protect species-at-risk and their habitat, which would be enforceable on land owned by the various levels of government, Aboriginal land-claims areas, and privately held property. Another crucial need is a comprehensive system to foster conservation-minded stewardship of critical habitats throughout the country.

DESIGNATION OF CONSERVATION STATUS

The Committee on the Status of Endangered Wildlife in Canada, as noted above, is responsible for designating the conservation status of native species[1] in Canada (Cook and Muir, 1984; Shank, 1999; COSEWIC, 2000). COSEWIC was created in 1977 by a joint decision of wildlife directors of the federal, provincial, and territorial governments, following a proposal developed with national non-governmental organizations (NGOs). COSEWIC includes academic and independent scientists and representatives of agencies of the federal government (i.e., Canadian Museum of Nature, Canadian Wildlife Service, Department of Fisheries and Oceans, and Parks Canada), the 10 provinces and three territories, and three non-governmental organizations (Canadian Nature Federation, Canadian Wildlife Federation, and World Wildlife Fund). Including the chairpersons of its eight subcommittees (or species specialist groups), COSEWIC has 28 voting members (as of mid-2000). However, the structure and role of COSEWIC have been dynamic over time and will change substantially in the year 2000 or soon afterward (for example, there will be a voting member to represent a new Aboriginal Community and Traditional Knowledge Subcommittee, and the three seats presently allocated to the Canadian Nature Federation, Canadian Wildlife Federation, and World Wildlife Fund will be opened up to NGOs more generally).

COSEWIC's eight subcommittees provide highly qualified expertise on particular taxonomic groups of organisms, with expert membership drawn mainly from museums, universities, and other institutions. The subcommittees are: (1) terrestrial mammals, (2) marine mammals, (3) birds, (4) freshwater fish, (5) marine fish, (6) reptiles and amphibians, (7) lepidoptera and molluscs, and (8) vascular plants, bryophytes, and lichens. Note that the subcommittees do not cover the entire breadth of Canadian biodiversity; COSEWIC does not yet have in-house expertise for designating the status of invertebrates (other than lepidoptera and molluscs), fungi, or micro-organisms. However, COSEWIC does have the flexibility to consider other groups of organisms by consulting outside experts.

The subcommittees develop lists of candidate species for consideration of conservation status, commission status reports, review those documents for accuracy and comprehensiveness, and propose a status for each species. The status reports are prepared according to a prescribed format and include information on the taxonomy, current and historical distribution and population size of the species, and threats and other limiting factors to its abundance in Canada and elsewhere. The subcommittees can also accept and review non-commissioned reports prepared by organizations or individuals interested in the potential designation of particular species. In 1999, for example, a status report on the conservation status of crayfish species in Canada was reviewed by COSEWIC (although COSEWIC had not commissioned that report). Status reports are periodically updated, preferably on about a 10-year basis.

Initially, COSEWIC had annual meetings, but since 1999 there have been two meetings each year. During the meetings new status reports and updates are reviewed and discussed. An attempt is made to reach a consensus when making a designation of the conservation status of a species. In the few cases where consensus is not possible, a two-thirds majority vote is used.

CATEGORIES OF CONSERVATION RISK

COSEWIC recognizes seven categories of risk, each of which has a specific meaning in terms of imminent threats to the future survival of a taxon (i.e., a species, subspecies, variety, or geographically distinct population). The rankings of endangered, threatened, and special concern are largely based on consideration of the population size of the taxon, recent trends in its abundance, and known and potential threats to its survival in Canada, including the extent and ecological integrity of its natural habitat. The categories of risk are the following:

- 'Extinct' refers to any species of wildlife that was formerly indigenous to Canada but no longer survives anywhere. Examples of extinct species whose range included Canada are the great auk (*Pinguinus impennis*), passenger pigeon (*Ectopistes migratorius*), Labrador duck (*Camptorhynchus labradorium*), sea mink (*Mustela macrodon*), deepwater cisco (*Coregonus johannae*), longjaw cisco (*Coregonus alpenae*), and eelgrass limpet (*Lottia alveus*). (See Appendix for a comprehensive list of taxa listed by COSEWIC.) As of May 2000, 12 Canadian species were extinct.
- 'Extirpated' refers to any species or subspecies that was formerly indigenous to Canada but now only survives in the wild elsewhere (usually in the neighbouring United States). Examples listed by COSEWIC include the black-footed ferret (*Mustela nigripes*), Atlantic grey whale (*Eschrichtius robustus*), greater prairie chicken (*Tympanuchus cupido*), pygmy short-horned lizard (*Phrynosoma douglassii douglassii*), paddlefish (*Polyodon spathula*), and blue-eyed mary (*Collinsia verna*). As of May 2000, 15 taxa were considered extirpated from Canada.
- 'Endangered' refers to indigenous taxa that are threatened with imminent extinction or extirpation throughout all or a significant portion of their Canadian range. Examples include the Vancouver Island marmot (*Marmota vancouverensis*), right whale (*Eubalaena glacialis*), whooping crane (*Grus americana*), burrowing owl (*Speotyto cunicularia*), piping plover (*Charadrius melodius*), blue racer snake (*Coluber constrictor*), thread-leaved sundew (*Drosera filiformis*), and seaside centipede lichen (*Heteroderma sitchensis*). As of May 2000, 102 taxa were considered endangered.
- 'Threatened' refers to any indigenous taxa that are likely to become endangered in Canada if factors affecting their vulnerability do not become reversed. Examples include the wood bison (*Bison bison athabascae*), sea otter (*Enhydra lutris*), marbled murrelet (*Brachyramphus marmoratus*), white-headed woodpecker (*Picoides albolarvatus*), massasauga rattlesnake (*Sistrurus catenatus*), and

American chestnut (*Castanea dentata*). As of May 2000, 71 taxa were considered threatened.

- 'Special concern' (formerly 'vulnerable') refers to any indigenous taxa that are not currently threatened but are at risk of becoming so because of characteristics that make them particularly sensitive to human activities or natural events. These include small or declining numbers, occurrence at the fringe of their range or in restricted areas, and habitat fragmentation. Examples include the grizzly bear (*Ursus arctos*), long-billed curlew (*Numenius americanus*), Pacific giant salamander (*Dicamptodon tenebrosus*), spotted turtle (*Clemmys guttata*), and prairie white-fringed orchid (*Platanthera leucophaea*). As of May 2000, 153 taxa were considered to be of special concern.

- 'Not at risk' refers to taxa that have been considered by COSEWIC and determined not currently to be at conservation risk in Canada. As of May 2000, 138 taxa had been reviewed and were considered not at risk.

- 'Data deficient' refers to taxa for which there is insufficient scientific information for the designation of conservation status. As of May 2000, 24 taxa were considered data deficient.

- Changes of status. The conservation status of a few species at risk has improved markedly in recent decades because of habitat protection, captive-breeding and release programs, and other enhancement activities. In a few other instances, better knowledge of the biology and ecology of the species has shown that its degree of risk was less than initially thought to be the case. These changes have resulted in the downlisting (i.e., the designation of a lesser status of conservation risk) or delisting (i.e., designation of a ranking of 'not at risk') of some species by COSEWIC. One such example is the downlisting of the *anatum* subspecies of the peregrine falcon (*Falco peregrinum anatum*) in 1999. The improved status of this species occurred because of decreased pollution by chlorinated hydrocarbons, and a population recovery program undertaken over a 25-year period by the Canadian Wildlife Service, the US Fish and Wildlife Service, World Wildlife Fund, and many other partners.

It must be recognized that the designation of species at risk is a continuing process and that changes of status occur regularly. As of May 2000, status reviews by COSEWIC had resulted in 44 taxa being uplisted (i.e., assigned a ranking of greater risk), 23 downlisted or delisted, while 167 were reassessed and warranted no change in their assigned status. From the recovery perspective, these data show that the status of the great majority of listed taxa has remained the same or is worsening.

Moreover, numerous taxa that are potentially at risk have not yet been investigated by COSEWIC. For instance, the conservation status of only a few species of invertebrates has been investigated, and species at risk in this group are greatly underrepresented in the COSEWIC list. More rapid progress by COSEWIC is constrained by a great shortage of funding for research and monitoring of species at risk in Canada and for the preparation of status reports. Another severe constraint is a lack of specialists

Table 2.1: Numbers of Taxa at Risk in Canada, by Major Category and Taxonomic Group

Taxonomic Group	Extinct	Extirpated	Endangered	Threatened	Special Concern	Total Listed	Recovery Plans*
Mammals	2	4	15	12	25	58	4
Birds	3	2	18	7	22	52	11
Reptiles	0	1	4	6	8	19	1
Amphibians	0	0	4	1	10	15	1
Fish	6	2	9	13	43	73	0
Invertebrates	1	4	6	1	2	14	0
Plants	0	2	44	31	40	117	0
Lichens	0	0	1	0	3	4	0

*Formally approved by RENEW.
SOURCES: World Wildlife Fund (1999); COSEWIC (2000).

with the necessary taxonomic and ecological knowledge to undertake work on species at risk. As of May 2000, COSEWIC had examined 515 taxa indigenous to Canada, of which 353 have been listed as being at risk (see Appendix). However, COSEWIC estimates that at least another 600 taxa require consideration of conservation risk.

It is also important to recognize that COSEWIC designations of conservation status have had no specific legal standing in Canada (although this will probably change for endangered and threatened taxa when federal legislation governing species at risk is passed by Parliament). Although legal consequences have not been inherent in COSEWIC designations, they have had a de facto importance in such planning processes as environmental impact assessment. Species listed as endangered, threatened, or of special concern by COSEWIC are routinely considered as VECs ('valued ecosystem components') in impact assessments, and potential consequences for their populations are considered.

Of course, COSEWIC is not the only agency that assigns designations of conservation status for Canadian species. Most of the provinces also have procedures for assigning rank for species occurring within their jurisdiction. At the international level, the US Fish and Wildlife Service has designated some species whose range includes Canada, while the World Conservation Union deals with global rankings. Of course, these other agencies are using different geographic criteria in their species assessments—provinces, for example, consider risk within their jurisdiction, and the US Fish and Wildlife Service does so for the United States or its regions.

MAJOR TAXONOMIC GROUPS OF THE LISTED SPECIES

Of the 353 taxa listed by COSEWIC in May 2000, 33 per cent are vascular plants, 21 per cent fish, 16 per cent mammals, 15 per cent birds, 6 per cent reptiles, 4 per cent amphibians, and 4 per cent invertebrates (Table 2.1). These frequencies of listings are substantially different from the relative proportions of these groups in the Canadian

Table 2.2.: Representation of Major Taxonomic Groups in the Biodiversity of Canada and in the COSEWIC Listings of Taxa at Risk

Group	Species in Canada			Listed by COSEWIC		
	Number Identified	Number Estimated	% of Total Estimated	Number	% of Total Listed	% of No. in Group
Viruses	200	150,000	53.2	0	0.0	0.0
Bacteria	2,400	23,200	8.2	0	0.0	0.0
Fungi	11,310	16,500	5.9	0	0.0	0.0
Protozoans	1,000	2,000	0.7	0	0.0	0.0
Algae	5,303	7,300	2.6	0	0.0	0.0
Lichens	2,500	2,800	1.0	4	1.1	0.1
Non-vascular plants	1,500	1,800	0.6	1	0.3	0.06
Vascular plants	4,153	4,400	1.6	117	33.2	2.7
Molluscs	1,500	1,635	0.6	8	2.3	0.5
Crustaceans	3,139	4,550	1.6	0	0.0	0.0
Arachnids	3,275	11,006	3.9	0	0.0	0.0
Insects	29,913	54,566	19.3	6	1.7	0.01
Fishes	1,100	1,600	0.6	73	20.7	4.6
Amphibians	42	44	0.02	15	4.2	34.1
Reptiles	42	42	0.02	19	5.4	45.2
Birds	430	430	0.2	52	14.7	12.1
Mammals	194	194	0.07	58	16.4	29.9
Total (rounded)	68,000	282,000		353		

NOTE: COSEWIC 'species' include some subspecies, varieties, and geographically defined populations.
SOURCES: Species in Canada: Environment Canada (1997); Freedman (2000). COSEWIC listings: World Wildlife Fund (1999); COSEWIC (2000).

milieu of biodiversity, in which viruses, insects, bacteria, and fungi comprise the largest fractions of species richness (accounting for 53, 19, 8, and 6 per cent, respectively, of the estimated number of species; Table 2.2). The disproportionate number of vertebrate animals and vascular plants listed by COSEWIC is partly related to their inherently high conservation risk, as many species in these groups are relatively large and maintain small populations. In addition, these groups are relatively well characterized biologically and ecologically, and they have a high profile with the public and even with most biologists. For these reasons, COSEWIC has focused its efforts on vertebrates and vascular plants, and has not yet effectively addressed the less well-known conservation needs of invertebrate animals and micro-organisms.

Of the taxonomic groups that COSEWIC does focus on, relatively large numbers of taxa of certain groups have been assigned conservation status. Of the 42 species of reptiles indigenous to Canada, 45 per cent are listed by COSEWIC, as are 34 per cent of the amphibians, 30 per cent of the mammals, and 12 per cent of the birds (Table 2.2). This disproportionate representation reflects the conservation reality of these groups, which contain numerous taxa that are sensitive to habitat loss and other eco-

Table 2.3: Distribution of Canadian Taxa at Risk among the Ecozones of Canada

Ecozone	% of Canada	Mammals	Birds	Reptiles	Amphibians	Fish	Plants
Marine Ecozones							
Pacific		6.8	0.0	4.0	0.0	2.6	0.0
Arctic Archipelago		4.5	0.0	0.0	0.0	2.6	0.0
Arctic Basin		0.0	0.0	0.0	0.0	0.0	0.0
Northwest Atlantic		12.4	0.0	0.0	0.0	1.3	0.0
Atlantic		6.8	0.0	4.0	0.0	1.3	0.0
Great Lakes		0.0	0.0	0.0	0.0	14.1	0.0
Terrestrial Ecozones							
Arctic Cordillera	2	2.2	0.0	0.0	0.0	0.0	0.0
Northern Arctic	14	3.3	2.7	0.0	0.0	1.3	0.0
Southern Arctic	8	2.2	5.3	0.0	0.0	0.0	0.0
Taiga Plains	6	4.5	3.5	0.0	0.0	0.0	0.0
Taiga Shield	15	5.6	3.5	0.0	0.0	0.0	0.8
Prairie	5	6.8	14.2	12.0	15.4	5.1	9.3
Atlantic Maritime	2	4.5	11.5	8.0	7.7	6.4	13.6
Mixed-wood Plains	2	4.5	18.6	48.0	38.4	29.4	53.4
Boreal Plains	7	2.2	3.5	0.0	0.0	0.0	0.0
Boreal Shield	20	6.8	13.3	12.0	0.0	9.0	8.5
Taiga Cordillera	3	2.2	2.7	0.0	0.0	0.0	0.0
Boreal Cordillera	5	4.5	0.0	0.0	0.0	1.3	0.0
Pacific Maritime	2	10.1	9.7	4.0	15.4	17.9	8.5
Montane Cordillera	5	9.0	8.8	8.0	23.1	7.7	5.9
Hudson Plains	4	1.1	2.7	0.0	0.0	0.0	0.0

NOTE: '% of Canada' refers to the amount of the land mass occupied by the ecozone; such data are not relevant for the marine ecozones. The other figures represent the percentages of listed taxa occurring in each ecozone. Some species occur in more than one ecozone.
SOURCE: Ecozone designation follows Ecological Stratification Working Group (1995).

logical changes associated with human activities, as well as the fact that their biology and ecology are relatively well known.

Overall, it appears that the conservation risk of indigenous taxa of vertebrates has been relatively well assessed in Canada. The risks of vascular plants are less well known, and those of non-vascular plants, lichens, invertebrates, and micro-organisms are virtually unstudied. This is a serious deficiency in our knowledge of risks to the indigenous biodiversity of Canada.

FACTORS INFLUENCING SPECIES AT RISK

Numerous biological, ecological, and anthropogenic factors interact to result in taxa becoming at risk of extinction. Taxa may be naturally rare and therefore sensitive to exploitation or habitat loss, as was apparently the case of the Labrador duck. Alternatively, they may be initially abundant yet decline in the face of relentless exploitation, as occurred with the great auk and passenger pigeon. Overall, however, anthropogenic stressors have been dominant in virtually all modern cases of endangerment and extinction. In this section, we briefly explore the most prominent factors influencing the conservation status of wildlife in Canada.

Ecozone Affinity

Most of the species at risk in Canada occur in the southern, most heavily settled parts of the country, where anthropogenic stressors are most intense. This region has the greatest density of human settlement and economic development, has suffered the most extensive losses of natural habitat, and experiences the most intense exposures to toxic chemicals and other human-caused stressors. In addition, many species reach the northern limit of their distribution in southern Canada and never occurred extensively in this country, resulting in an intrinsic vulnerability to the effects of habitat loss and population decline.

The highest frequency of species at risk (30 per cent; Table 2.3) in the terrestrial ecozones is in the Mixed-wood Plains of the southernmost areas of Ontario and Quebec, a region where almost all of the natural habitat has been converted into agricultural and urbanized land uses. This ecozone comprises only 2 per cent of the land area of Canada, but sustains 30 per cent of the taxa listed by COSEWIC. The next highest frequencies (range of 9–13 per cent) are in other southern ecozones where anthropogenic stressors are intense and extensive: the Pacific Maritime (13 per cent), Atlantic Maritime (11 per cent), Boreal Shield (11 per cent), Prairie (10 per cent), and Montane Cordillera (9 per cent). The other nine ecozones of Canada, comprising about two-thirds of the land area, account for only 16 per cent of the taxa at risk. (For a map showing terrestrial and marine ecozones of Canada, see Figure 3.2 in Chapter 3, below.)

Similarly, about 26 per cent of the marine taxa at risk occur in the Great Lakes, while 22 per cent are in the Pacific Marine ecozone, 22 per cent in the Northwest Atlantic, and 19 per cent in the Atlantic (Table 2.3). Only 6 per cent of the marine taxa at risk occur in the two Arctic marine ecozones.

Habitat Affinity

We scored taxa listed by COSEWIC on the basis of their habitat use. Habitat types are extremely diverse, but for simplicity the categories presented here are: tundra (arctic or alpine), prairie/grassland, semi-desert, boreal forest, montane forest, temperate hardwood forest, coastal rain forest, freshwater wetland (swamp, marsh, bog, fen), lentic (lake and pond), lotic (river and stream), estuary, marine coastal, and marine open water.

About 31 per cent of the listed mammals utilize marine open water habitats; these are mainly whales (Table 2.4). Other habitats used relatively frequently (7–11 per cent) by listed mammals are: prairie/grassland, boreal forest, temperate rain forest, and montane forest.

The most frequent habitat affinities of listed birds are prairie/grassland (23 per cent) and temperate forest (21 per cent; most of these are species of southern hardwood forest of southern Ontario). Other relatively frequently utilized habitats (5–12

Table 2.4: Distribution of Canadian Taxa at Risk among Major Habitat Types

Habitat	Mammals	Birds	Reptiles	Amphibians	Fish	Plants
Arctic tundra	4.2	4.8	0.0	0.0	0.0	1.5
Alpine tundra	1.4	2.4	0.0	0.0	0.0	0.0
Boreal forest	11.1	6.0	0.0	0.0	0.0	2.3
Montane forest	6.9	4.8	4.3	12.0	0.0	2.3
Temperate forest	4.2	20.6	43.6	12.0	0.0	35.8
Temperate rain forest	9.7	4.8	4.3	8.0	0.0	1.5
Prairie/grassland	11.1	22.9	17.4	16.0	0.0	17.5
Semi-desert	1.4	1.2	8.7	4.0	0.0	0.0
Wetland	1.4	7.2	0.0	0.0	0.0	20.6
Lentic	0.0	2.4	17.4	44.0	50.0	9.2
Lotic	2.8	1.2	0.0	4.0	42.6	3.1
Marine open water	30.5	7.2	4.3	0.0	3.7	0.0
Estuary	0.0	2.4	0.0	0.0	3.7	0.8
Coastal	4.2	12.1	0.0	0.0	0.0	2.3

Numbers are percentages of listed taxa occurring within the habitat. Some species occur in more than one habitat.

per cent) are marine coastal and open water, freshwater wetland, boreal and montane forests, and temperate rain forest.

The most frequently utilized habitat of listed reptiles is temperate forest (44 per cent; these are mostly species of southern Ontario). Other important habitats (9–17 per cent) are semi-desert, prairie/grassland, and lakes and ponds. Listed amphibians breed mostly in lakes and ponds, within landscapes dominated by temperate or montane forest or prairie/grassland. Listed fish mostly occur in lentic (50 per cent) and lotic (43 per cent) freshwaters, with only 4 per cent occurring in estuaries and 4 per cent in marine habitats.

Listed plants occur most frequently in temperate forest (36 per cent; most of these are species of the eastern deciduous forest, including the Carolinian zone). Other important habitats are wetlands (21 per cent) and prairie/grassland (18 per cent).

Anthropogenic Influences

Humans affect indigenous species in diverse ways, but the most important classes of influences are degradation and loss of habitat, toxic chemicals, harvesting, and the ecological effects of introduced species. In almost all cases, species at risk are simultaneously affected by several of these anthropogenic stressors.

Degradation or loss of habitat is by far the most important stressor affecting taxa at risk in Canada (Table 2.5). This factor is cited as being important to 100 per cent of the listed reptiles, amphibians, invertebrates, and lichens, and to 99 per cent of the plants, 90 per cent of the birds, 85 per cent of the fish, and 67 per cent of the mammals.

Exploitation by humans is an important stressor for those species that are hunted, fished, collected as pets, or economically valuable in other ways. Exploitation by

Table 2.5: Classes of Anthropogenic Influences Causing Stress to Species Listed by COSEWIC

Taxonomic Group	Class of Anthropogenic Influence (%)				Number of Listed Taxa
	Habitat Loss	Non-Native Species	Toxic Chemicals	Exploitation	
Mammals	67	0	29	62	55
Birds	90	10	33	33	52
Reptiles	100	6	11	47	19
Amphibians	100	25	33	25	12
Invertebrates	100	0	46	8	13
Fish	85	17	33	19	72
Plants	99	16	5	7	112
Lichens	100	0	25	0	4

NOTE: Data are percentages of listed taxa affected by each class of stressor. Sums exceed 100% because most species are significantly affected by more than one stressor.
SOURCE: COSEWIC status reports summarized in World Wildlife Fund (1999).

humans is considered an important cause of mortality for 62 per cent of the listed mammals, 42 per cent of the reptiles, 33 per cent of the birds, and 25 per cent of the amphibians (Table 2.5).

Habitat degradation, competition, or predation associated with invasive non-native species is considered an important stressor to 25 per cent of listed amphibians, 17 per cent of fish, 10 per cent of birds, and 16 per cent of plants (Table 2.5). The non-native species affect animals at risk mostly through intense competition or predation, but they largely affect listed plants by degrading the suitability of their habitat.

Many listed species are thought to be significantly stressed by exposure to toxic chemicals in their environment (such as pesticides, toxic gases, and ionic metals). This factor is considered an important stressor for 46 per cent of the listed invertebrates, and for 25–33 per cent of the amphibians, birds, fish, and mammals. In almost all of these cases, however, habitat loss and degradation are considerably more important stressors than exposure to toxic substances.

RECOVERY PLANS FOR LISTED TAXA

Once a species is listed by COSEWIC as endangered or threatened in Canada, a parallel body known as RENEW (Recovery of Nationally Endangered Wildlife) is mandated to prepare a plan that would ensure the recovery of its population to a safer level of abundance (RENEW, 2000). As of 2000, recovery plans had been approved for 18 taxa (all are vertebrate animals), while another 24 were in various stages of preparation, and for one taxon (black-footed ferret) actions were on hold. In addition, a recovery plan is being prepared for one ecosystem at risk (semi-desert in the South Okanagan of British Columbia, which provides habitat for various species at risk). Recovery planning to date by RENEW involves endangered and threatened terrestrial vertebrates. Of the 18 taxa for which RENEW recovery plans were in place in 2000, 12 are for birds, four are for mammals, and one each a reptile and an amphibian. In addition, several recovery plans have been prepared outside of the RENEW process; for example, World Wildlife Fund has prepared one for the northern right whale in partnership with the Department of Fisheries and Oceans, and another for tall-grass prairie in southern Ontario in partnership with the Ontario Ministry of Natural Resources.

ECOSYSTEMS AT RISK

Some of the natural ecosystems of Canada are also at great risk of being lost because they occur only as small, possibly unsustainable remnants of their former extent. The most critically at-risk indigenous ecosystems of Canada include:

- tall-grass prairie of southwestern Ontario and southeastern Manitoba;
- Carolinian (southern deciduous) forest of southern Ontario;
- dry Douglas fir and Garry oak forest of southwestern coastal British Columbia;
- semi-desert of southeastern British Columbia;
- various types of old-growth forest in all parts of forested Canada;
- natural wetlands throughout most of southern Canada. (Freedman, 2000)

Most of these imperilled ecosystems are also rich in species at risk. It is crucial that the remaining areas of these ecosystems be preserved in parks, ecological reserves, and other kinds of protected areas.

OTHER ACTIONS TO PROTECT SPECIES AT RISK IN CANADA

It is not, of course, sufficient merely to designate species as being at risk of extirpation or extinction. If their status in Canada is to be improved, the species and their habitat must also be protected. All major stresses on the species must clearly be reduced significantly or removed. The federal government has not yet enacted legislation to protect species at risk; however, that may change in the near future.

In late 1996, the government of Canada tabled an Endangered Species Act, with the aim of providing protection for species occurring on federal lands. That legislation was not passed, however, as Parliament recessed before the proposed Act had been considered. A substantially modified version of that legislation, called the Species at Risk Act (SARA), was tabled in Parliament in April 2000, but it, too, died when an election was called. Potentially, SARA provided the Minister of Environment with the authority to prohibit the destruction of endangered or threatened species and their critical habitat on all lands in Canada, although the broader intent was to work in a collaborative manner with provincial, territorial, and other levels of government. The proposed legislation would legally have recognized COSEWIC and would have provided for rigorous, independent, and public scientific assessments of the status of species potentially at risk (although, according to this legislative proposal, the federal cabinet, not COSEWIC, would make the final decision about legal listing and protection). SARA also proposed incentives for conservation and stewardship measures, as well as provisions for compensating private land-owners in extraordinary cases of loss of livelihood or other extreme opportunity costs associated with the presence of species at risk or their critical habitat. SARA remained extremely controversial, however, largely because it did not contain sufficiently effective provisions to ensure the protection of critical habitat.

Of course, the protection of critical habitat is not only a federal responsibility— this mandate must also be effectively addressed by legislation enacted by all provinces and territories. Species-at-risk legislation already exists in Manitoba, New Brunswick, Nova Scotia, Ontario, and Quebec, and provides varying levels of protection to species within those jurisdictions. Collectively, however, existing Canadian legislation does not provide sufficient protection for species at risk and their critical habitats. It is essential that an effective and co-ordinated system is developed for the protection of species at risk and their habitats in Canada; a piecemeal approach would result in uneven levels of protection and would be insufficient to conserve indigenous biodiversity.

The development of legislation to protect species at risk is an ongoing undertaking in Canada. Governments feel the need to demonstrate that they are making rapid progress towards economic and environmental sustainability, and a key component of that involves the protection of species at risk and their habitats. Important progress has been made in this direction by Canadian governments, NGOs, and other partners in conservation. Nevertheless, actions to protect species at risk have been significantly lacking in important respects (particularly in regard to the protection of critical habitat), and much work remains to be done to ensure their conservation and recovery in Canada. Appropriate changes to the federal, provincial, and territorial initiatives may result from the lobbying efforts of Canadian NGOs (under the umbrella of the Canadian Endangered Species Campaign) and of hundreds of scientists signatory to several widely subscribed petitions to the federal government (Centre for Biodiversity Research, 1999).

Most of the critical habitat in Canada occurs in relatively southern parts of the country, and much of it is on privately owned land. An extremely useful development since the mid-1990s has been the passage by the federal government and most provinces of legislation allowing private individuals and corporations to be given a measure of tax relief in exchange for their donation of conservation easements to designated environmental organizations (such as the Nature Conservancy of Canada and provincial land trusts). Under such an arrangement, the conservation easement (also known as a covenant) allows the landowner to retain title to the property, but 'ownership' of the natural habitat belongs to the environmental organization. This arrangement means that inappropriate land-use activities, such as conversion of the property to agricultural or residential usage, can be prohibited. Up to late 1999, the Nature Conservancy of Canada had negotiated 51 conservation easements, involving about 10,600 hectares of natural land (Nature Conservancy of Canada, pers. comm.).

In addition, Canada is signatory to the Convention on International Trade in Endangered Species (CITES), an international agreement that commits nations to controlling or preventing the trade in endangered species. Canada's responsibilities under CITES are to monitor and report on its international trade in all species listed in the 'red books' prepared by the World Conservation Monitoring Centre (trade in about 400 species of animals and 150 plants is prohibited by CITES; trade in about 2,000 species of animals and 60,000 plants is monitored). Canadian species listed by CITES include 21 species of mammals, 10 birds, four reptiles or amphibians, two fish, all orchids, and four other plants (Freedman, 2000). By providing a legal system of regulation for the export of species at risk and their body parts, CITES allows the federal government to regulate or prevent that trade. However, some trade within Canada may still be considered legal, and the degree of enforcement of the CITES provisions is not sufficient to prevent all illegal poaching and international trade.

In addition to legislation, there are other critical tactics in the 'recovery tool kit' for species at risk in Canada. A major need is to increase the management advice and financial incentives available to landowners in support of conservation stewardship on private land. This element is being advanced in Canada by provincial and federal governments and by NGOs such as World Wildlife Fund, the Nature Conservancy of Canada, Ducks Unlimited, and Wildlife Habitat Canada. However, these efforts are not sufficiently funded and must be expanded considerably if they are to be truly effective.

One final element that must be addressed is the need for an improved level of environmental literacy in Canada, including public knowledge of the importance of indigenous biodiversity, its conservation needs, and its fundamental role in maintaining the healthy ecosystems on which we all depend. If all Canadians were well informed about this issue, they might be more supportive of the substantial level of spending and other actions needed to deal with the risks to native species and ecosystems. Increasing the level of environmental literacy is a key goal of many non-governmental organizations and governments in Canada, and much progress has been made. Nevertheless, much more activity is needed in this field, so that the protection of Canadian biodiversity can achieve the high public and political profile it deserves.

NOTE

1. In this context, 'species' is defined as any indigenous species, subspecies, variety, or geographically defined population of wild fauna and flora; this sense is also used throughout this chapter, interchangeably with the word 'taxon'.

REFERENCES

Centre for Biodiversity Research. 1999. '640 scientists recommend federal government take action on endangered species protection', University of British Columbia. http://www.zoology.ubc.ca/~otto/biodiversity/

Cook, F.R., and D. Muir. 1984. 'The Committee on the Status of Endangered Wildlife in Canada (COSEWIC): History and Progress', Canadian Field-Naturalist 98: 63–70.

COSEWIC. 2000. Canadian Species At Risk. Ottawa: Committee on the Status of Wildlife in Canada, May.

Ecological Stratification Working Group. 1995. *A National Ecological Framework for Canada.* Ottawa: Environment Canada.

Environment Canada. 1997. *The State of Canada's Environment.* State of the Environment Reporting Organization. Ottawa.

——. 2000. *RENEW Report No. 10, 1999–2000.* Ottawa: Canadian Wildlife Service.

——. 2000. *The Green Lane: COSEWIC.* http://www.cosewic.gc.ca/COSEWIC

Freedman, B. 2000. *Environmental Science: A Canadian Perspective,* 2nd edn. Toronto: Prentice-Hall Canada.

Shank, C.C. 1999. 'The Committee on the Status of Endangered Wildlife in Canada (COSEWIC): A 21-Year Retrospective', *Canadian Field-Naturalist* 113: 318–41.

World Wildlife Fund. 1999. *List of Canadian Wildlife At Risk.* Toronto: World Wildlife Fund (Canada).

APPENDIX: TAXA LISTED AS BEING AT RISK IN CANADA (MAY 2000)

Lichens
Heterodermia sitchensis; Seaside Centipede; endangered, BC
Hypogymnia heterophylla; Seaside Bone; special concern, BC
Nephroma occultum; Cryptic Paw; special concern, BC
Pseudocyphellaria ranierensis; Oldgrowth Specklebelly; special concern, BC

Mosses
Bartramia stricta; Apple Moss; endangered, BC

Plants
Abronia micrantha; Sand Verbena; threatened, Alta, Sask.
Achillea millefolium var. megacephalum; Large-headed Wooly Yarrow; special concern, Sask.
Adiantum capillus-veneris; Southern Maidenhair Fern; endangered, BC
Agalinis gattingeri; Gattinger's Agalinis; endangered, Ont.
Agalinis skinneriana; Skinner's Agalinis; endangered, Ont.
Aletris farinosa; Colicroot; threatened, Ont.
Ammania robusta; Scarlet Ammania; endangered, BC, Ont.
Arisaema dracontium; Green Dragon; special concern, Ont.
Armeria maritima ssp. *interior*; Athabasca Thrift; special concern, Sask.
Aster anticostensis; Anticosti Aster; threatened, NB, Que.
Aster curtus; White-Top Aster; threatened, BC
Aster divaricatus; White Wood Aster; threatened, Ont., Que.
Aster laurentianus; Gulf of St Lawrence Aster; special concern, NB, PEI, Que.
Aster praealtus; Willow Aster; special concern, Ont.
Aster prenanthoides; Crooked-Stem Aster; special concern, Ont.
Aster subulatus var. *obtusifolius*; Bathurst Aster; special concern, NB
Astragalus robbinsii var. fernaldii; Fernald's Milk-Vetch; special concern, Nfld, Que.

Azolla mexicana; Mosquito Fern; threatened, BC
Balsamorhiza deltoidea; Deltoid Balsamroot; endangered, BC
Bartonia paniculata; Branched Bartonia; special concern, Ont.
Braya fernaldii; Fernald's Braya; threatened, Nfld
Braya longii; Long's Braya; endangered, Nfld
Buchloe dactyloides; Buffalograss; special concern, Man., Sask.
Buchnera americana; Bluehearts; endangered, Ont.
Cacalia plantaginea; Indian Plantain; special concern, Ont.
Camassia scilloides; Wild Hyacinth; special concern, Ont.
Carex juniperorum; Juniper Sedge; endangered, Ont.
Carex lupuliformis; False Hop Sedge; endangered, Ont., Que.
Castanea dentata; American Chestnut; threatened, Ont.
Castilleja levisecta; Golden Paintbrush; endangered, BC
Celtis tenuifolia; Dwarf Hackberry; special concern, Ont.
Cephalanthera austinae; Phantom Orchid; threatened, BC
Chenopodium subglabrum; Smooth Goosefoot; special concern, Alta, Man., Sask.
Chimaphila maculata; Spotted Wintergreen; endangered, Ont.
Cicuta maculata var. *victorinii*; Victorin's Water Hemlock; special concern, Que.
Cirsium pitcheri; Pitcher's Thistle; endangered, Ont.
Clethra alnifolia; Sweet Pepperbush; threatened, NS
Collinsia verna; Blue-Eyed Mary; extirpated, Ont.
Coreopsis rosea; Pink Coreopsis; endangered, NS
Cryptantha minima; Tiny Cryptanthe; endangered, Alta, Sask.
Cypripedium candidum; Small White Lady's-Slipper; endangered, Ont., Man.
Dalea villosa var. villosa; Hairy Prairie-Clover; threatened, Man., Sask.
Deschampsia mackenzieana; Mackenzie Hairgrass; special concern, Sask.
Desmodium illinoense; Illinois Tick-Trefoil; extirpated, Ont.
Drosera filiformis; Thread-Leaved Sundew; endangered, NS
Dryopteris arguta; Coastal Wood Fern; special concern, BC
Eleocharis tuberculosa; Tubercled Spike-rush; threatened, NS
Epipactis gigantea; Giant Helleborine; special concern, BC
Erigeron philadelphicus ssp. *provancheri*; Provancher's Fleabane; special concern, Que.
Frasera caroliniensis; American Columbo; special concern, Ont.
Fraxinus quadrangulata; Blue Ash; threatened, Ont.
Gentiana alba; White Prairie Gentian; endangered, Ont.
Gentiana victorinii; Victorin's Gentian; special concern, Que.
Geum peckii; Eastern Mountain Avens; endangered, NS
Gymnocladus dioica; Kentucky Coffee Tree; threatened, Ont.
Halimolobos virgata; Slender Mouse-Ear-Cress; threatened, Alta, Sask.
Hibiscus moscheutos; Swamp Rose Mallow; special concern, Ont.
Hydrastis canadensis; Golden Seal; threatened, Ont.
Hydrocotyle umbellata; Water-Pennywort; threatened, NS
Iris missouriensis; Western Blue Flag; threatened, Alta

Isoetes bolanderi; Bolander's Quillwort; special concern, Alta

Isoetes engelmannii; Engelmann's Quillwort; endangered, Ont.

Isopyrum biternatum; False Rue-Anemone; special concern, Ont.

Isotria medeoloides; Small Whorled Pogonia; endangered, Ont.

Isotria verticillata; Large Whorled Pogonia; endangered, Ont.

Juncus caesariensis; New Jersey Rush; special concern, NS

Justicia americana; American Water-Willow; threatened, Ont., Que.

Lachnanthes caroliana; Redroot; threatened, NS

Lespedeza virginica; Slender Bush Clover; endangered, Ont.

Liatris spicata; Dense Blazing Star; special concern, Ont.

Lilaeopsis chinensis; Lilaeopsis; special concern, NS

Limnanthes macounii; Macoun's Meadowfoam; special concern, BC

Liparis liliifolia; Purple Twayblade; endangered, Ont.

Lipocarpha micrantha; Small-Flowered Lipocarpha; threatened, BC, Ont.

Lophiola aurea; Golden Crest; threatened, NS

Lotus formosissimus; Seaside Birds-Foot Lotus; endangered, BC

Lupinus lepidus var. *lepidus*; Prairie Lupine; endangered, BC

Magnolia acuminata; Cucumber Tree; endangered, Ont.

Morus rubra; Red Mulberry; endangered, Ont.

Opuntia humifusa; Eastern Prickly Pear Cactus; endangered, Ont.

Oxytropis lagopus; Hare-Footed Locoweed; special concern, Alta

Panax quinquefolium; American Ginseng; endangered, Ont., Que.

Pedicularis furbishiae; Furbish's Lousewort; endangered, BC

Phegopteris hexagonoptera; Broad Beech Fern; special concern, Ont., Que.

Plantago cordata; Heart-Leaved Plantain; endangered, Ont.

Platanthera leucophaea; Eastern Prairie White Fringed Orchid; special concern, Ont.

Platanthera praeclara; Western Prairie White Fringed Orchid; endangered, Man.

Polemonium van-bruntiae; van Brunt's Jacob's Ladder; threatened, Que.

Polygala incarnata; Pink Milkwort; endangered, Ont.

Potamogeton hillii; Hill's Pondweed; special concern, Ont.

Ptelea trifoliata; Hop Tree; special concern, Ont., Que.

Pycnanthemum incanum; Hoary Mountain Mint; endangered, Ont.

Quercus shumardii; Shumard Oak; special concern, Ont.

Ranunculus alismaefolius var. *alismaefolius*; Water-Plantain Buttercup; endangered, BC

Rosa setigera; Climbing Prairie Rose; special concern, Ont.

Rotala ramosior; Toothcup; endangered, BC, Ont.

Sabatia kennedyana; Plymouth Gentian; threatened, NS

Salix brachycarpa var. *psammophila*; Sand-dune Short-capsuled Willow; special concern, Sask.

Salix silicicola; Felt-leaf Willow; special concern, Sask., Nunavut

Salix turnorii; Tumor's Willow; special concern, Sask.

Scirpus longii; Long's Bulrush; special concern, NS

Smilax rotundifolia; Round-Leaved Greenbrier; threatened, Ont. (Ontario population)

Solidago speciosa var. *rigidu scula*; Showy Goldenrod; endangered, Ont.

Stylophorum diphyllum; Wood Poppy; endangered, Ont.

Symphyotrichum sericeum (Virgulus sericeus); Western Silver-leaved Aster; threatened, Man., Ont.

Tanacetum huronense var. *floccosum*; Floccose Tansy; special concern, Sask.

Tephrosia virginiana; Goat's-Rue; endangered, Ont.

Tradescantia occidentalis; Western Spiderwort; threatened, Alta, Man., Sask.

Trichophorum (Scirpus) verecundus; Few-Flowered Club-Rush; endangered, Ont.

Trillium flexipes; Drooping Trillium; endangered, Ont.

Triphora trianthophora; Nodding Pogonia; endangered, Ont.

Triphysaria versicolor versicolor; Bearded Owl Clover; endangered, Ont.

Vaccinium stamineum; Deerberry; threatened, Ont.

Virgulus sericeus; Western Silver-Leaf Aster; special concern, Man., Ont.

Viola pedata; Bird's-Foot Violet; threatened, Ont.

Viola praemorsa ssp. *praemorsa*; Yellow Montane Violet; threatened, BC

Woodsia obtusa; Blunt-Lobed Woodsia; endangered, Ont., Que.

Yucca glauca; Soapweed; threatened, Alta

Molluscs

Alasmidontia heterodon; Dwarf Wedgemussel; extirpated, NB

Eploblasma tortulosa rangiana; Northern Riffleshell; endangered, Ont.

Halliotis kamtschatkana; Northern Abalone; threatened, BC

Lampsilis fasciola; Wavy-Rayed Lampmussel; endangered, Ont.

Lottia alveus; Eelgrass Limpet; extinct, Nfld, NS, Que.

Physella johnsoni; Banff Springs Snail; endangered, Alta

Physella wrighti; Hotwater Physa; endangered, BC

Villosa fabalis; Rayed Bean; endangered, Ont.

Insects

Caollophrys irus; Froseted Elfin Butterfly; extirpated, Ont.

Coenonympha tullia nipisiquit; Maritime Ringlet Butterfly; endangered, NB, Que.

Danaus plexippus; Monarch Butterfly; special concern, all provinces

Euchloe ausonides; Island Marble Butterfly; extirpated, BC

Limenitis weidemeyerii; Weidemeyer's Admiral; special concern, Alta

Lycaeides melissa samuelis; Karner Blue Butterfly; extirpated, Ont.

Fish

Acantholumpenus mackayi; Blackline Prickleback; special concern, NWT, Nunavut (Arctic Ocean)

Anarhichas orientalis; Bering Wolffish; special concern, NWT, Nunavut (Arctic Ocean)

Acipenser brevirostrum; Shortnose Sturgeon; special concern, NB
Acipenser medirostris; Green Sturgeon; special concern, BC
Acipenser transmontanus; White Sturgeon; special concern, BC
Ammocrypta pellucida; Eastern Sand Darter; threatened, Ont., Que.
Catostomus species; Salish Sucker; endangered, BC
Clinostomus elongatus; Redside Dace; special concern, Ont.
Coregonus alpenae; Longjaw Cisco; extinct, Ont.
Coregonus clupeaformis; Lake Simcoe Whitefish; threatened, Ont.
Coregonus huntsmani; Atlantic Whitefish; endangered, NS
Coregonus johannae; Deepwater Cisco; extinct, Ont.
Coregonus kiyi; Kiyi; special concern, Ont.
Coregonus nigripinnis; Blackfin Cisco; threatened, Ont.
Coregonus reighardi; Shortnose Cisco; threatened, Ont.
Coregonus zenithicus; Shortjaw Cisco; threatened, Alta, Man., Ont., Sask.
Coregonus species; Spring Cisco; special concern, Que.
Coregonus species; Squanga Whitefish; special concern, Yukon
Cottus bairdi ssp. *hubbsi*; Columbia Mottled Sculpin; special concern, BC
Cottus confusus; Shorthead Sculpin; threatened, BC
Cottus species; Cultus Pygmy Sculpin; special concern, BC
Erimystax x-punctatus; Gravel Chub; extirpated, Ont.
Erimyzon sucetta; Lake Chubsucker; special concern, Ont.
Etheostoma blennioides; Greenside Darter; special concern, Ont.
Fundulus diaphanus; Banded Killifish; special concern, Nfld population
Fundulus notatus; Blackstripe Topminnow; special concern, Ont.
Gadus morhua; Atlantic Cod; special concern, Atlantic Ocean
Gasterosteus aculeatus; Charlotte Unarmoured Stickleback; special concern, BC
Gasterosteus sp.; Benthic Hadley Lake Stickleback; extinct, BC
Gasterosteus sp.; Benthic Paxton Lake Stickleback; endangered, BC
Gasterosteus sp.; Benthic Texada Island Stickleback; threatened, BC
Gasterosteus sp.; Benthic Vananda Creek Stickleback; endangered, BC
Gasterosteus sp.; Enos Lake Stickleback; threatened, BC
Gasterosteus sp.; Giant Stickleback; special concern, BC
Gasterosteus sp.; Limnetic Hadley Lake Stickleback; extinct, BC
Gasterosteus sp.; Limnetic Texada Island Stickleback; threatened, BC
Gasterosteus sp.; Limnetic Paxton Lake Stickleback; endangered, BC
Gasterosteus sp.; Limnetic Vananda Creek Stickleback; endangered, BC
Hybognathus argyritus; Western Silvery Minnow; special concern, Alta
Ichthyomyzon castaneus; Chestnut Lamprey; special concern, Ont.
Ichthyomyzon fossor; Northern Brook Lamprey; special concern, Man., Ont., Que.
Ictiobus cyprinellus; Buffalo Bigmouth; special concern, Man., Ont., Sask.
Ictiobus niger; Black Buffalo; special concern, Ont.
Lampetra macrostoma; Lake Lamprey; special concern, BC
Lampetra richardsoni; Morrison Creek Lamprey; endangered, BC

Lepisosteus oculatus; Spotted Gar; special concern, Ont.
Lepomis auritus; Redbreast Sunfish; special concern, NB
Lepomis gulosus; Warmouth; special concern, Ont.
Lepomis humilis; Orange-Spotted Sunfish; special concern, Ont.
Macrhybopsis storeriana; Silver Chub; special concern, Man., Ont.
Minytrema melanops; Spotted Sucker; special concern, Ont.
Moxostoma carinatum; River Redhorse; special concern, Ont., Que.
Moxostoma duquesnei; Black Redhorse; threatened, Ont.
Moxostoma hubbsi; Copper Redhorse; threatened, Que.
Myoxocephalus quadricornis; Fourhorn Sculpin; special concern, NWT, Nunavut
   (Arctic Islands, freshwater form)
Myoxocephalus thompsoni; Deepwater Sculpin; threatened, Ont. (Great Lakes popula-
   tion)
Notropis anogenus; Pugnose Shiner; special concern, Ont.
Notropis bifrenatus; Bridle Shiner; special concern, Ont.
Notropis dorsalis; Bigmouth Shiner; special concern, Man.
Notropis photogenis; Silver Shiner; special concern, Ont.
Notropis rubellus; Rosyface Shiner; special concern, Man. population
Noturus insignis; Margined Madtom; threatened, Ont., Que.
Noturus miurus; Brindled Madtom; special concern, Ont.
Noturus stigmosus; Northern Madtom; special concern, Ont.
Opsopoeodus emiliae; Pugnose Minnow; special concern, Ont.
Osmerus species; Lake Utopia Dwarf Smelt; threatened, NB
Percina copelandi; Channel Darter; threatened, Ont., Que.
Polyodon spathula; Paddlefish; extirpated, Ont.
Rhinichthys cataractae smithi; Banff Longnose Dace; extinct, Alta
Rhinichthys osculus; Speckled Dace; special concern, BC
Rhinichthys umatilla; Umatilla Dace; special concern, BC
Rhinichthys species; Nooksack Dace; endangered, BC
Salvelinus fontinalis timagamiensis; Aurora Trout; endangered, Ont.
Sardinops sagax; Pacific Sardine; special concern, Pacific Ocean
Stizostedion vitreum glaucum; Blue Walleye; extinct, Ont.

Amphibians
Acris crepitans; Northern Cricket Frog; endangered, Ont.
Ambystoma texanum; Smallmouth Salamander; special concern, Ont.
Ascaphus truei; Tailed Frog; endangered, BC (Southern Mountain population)
Ascaphus truei; Tailed Frog; special concern, BC (Pacific Coast population)
Bufo cognatus; Great Plains Toad; special concern, Alta, Man., Sask.
Bufo fowleri; Fowler's Toad; threatened, Ont.
Desmognathus ochrophaeus; Mountain Dusky Salamander; special concern, Que.
Dicamptodon tennebrosus; Pacific Giant Salamander; special concern, BC
Gyrinophallus porphyriticus; Spring Salamander; special concern, Que.

Plethodon idahoensis; Coeur d'Alene Salamander; special concern, BC
Rana aurora; Northern Red-Legged Frog; special concern, BC
Rana pipiens; Northern Leopard Frog; endangered, BC (S. Rocky Mountain population)
Rana pipiens; Northern Leopard Frog; special concern, Alta, Man., Sask. (Prairie population)
Rana pretiosa; Oregon Spotted Frog; endangered, BC
Spea intermontana; Great Basin Spadefoot Toad; special concern, BC

Reptiles
Apalone spinifera; Spiny Softshell Turtle; threatened, Ont., Que.
Clemmys guttata; Spotted Turtle; special concern, Ont., Que.
Clemmys insculpta; Wood Turtle; special concern, NB, NS, Ont., Que.
Coluber constrictor foxii; Blue Racer; endangered, Ont.
Coluber constrictor flaviventris; Eastern Yellow-Bellied Racer; special concern, Sask.
Contia tenuis; Sharp-Tailed Snake; endangered, BC
Dermochelys coriacea; Leatherback Turtle; endangered, Atlantic and Pacific Oceans
Elaphe obsoleta obsoleta; Black Rat Snake; threatened, Ont.
Elaphe vulpina gloydi; Eastern Fox Snake; threatened, Ont.
Emydoidea blandingi; Blanding's Turtle; threatened, NS population
Eumeces fasciatus; Five-Lined Skink; special concern, Ont.
Eumeces septentrionalis septentrionalis; Northern Prairie Skink; special concern, Man.
Heterodon platirhinos; Eastern Hognose Snake; special concern, Ont.
Nerodia sipedon insularum; Lake Erie Water Snake; endangered, Ont.
Phrynosoma douglassii brevirostre; Eastern Short-Horned Lizard; special concern, Alta, Sask.
Phrynosoma douglassii douglassii; Pygmy Short-Horned Lizard; extirpated, BC (BC population)
Regina septemvittata; Queen Snake; threatened, Ont.
Sistrurus catenatus catenatus; Eastern Massasauga Rattlesnake; threatened, Ont.
Thamnophis butleri; Butler's Garter Snake; special concern, Ont.

Birds
Accipiter gentilis laingi; Queen Charlotte Goshawk; special concern, BC
Ammodramus henslowii; Henlow's Sparrow; endangered, Ont.
Anthus spragueli; Sprague's Pipit; threatened, Alta, Man., Sask.
Ardea herodias fannini; Pacific Great Blue Heron; special concern, BC
Asio flammeus; Short-Eared Owl; special concern, all provinces and territories
Brachyramphus marmoratus; Marbled Murrelet; threatened, BC
Buteo lineatus; Red-Shouldered Hawk; special concern, Ont., Que.
Buteo regalis; Ferruginous Hawk; special concern, Alta, Man., Sask.
Camptorhynchus labradorius; Labrador Duck; extinct, NB, Nfld, NS, Que.
Catharus bicknelii; Bicknell's Thrush; special concern, NB, NS, Que.

Centrocercus urophasianus phaios; Sage Grouse; extirpated, BC

Centrocercus urophasianus urophasianus; Sage Grouse; endangered, Alta, Sask.

Charadrius melodus; Piping Plover; endangered, Alta, Man., NB, Nfld, NS, Ont., PEI, Que., Sask.

Charadrius montanus; Mountain Plover; endangered, Alta, Sask.

Colinus virginianus; Northern Bobwhite; endangered, Ont.

Coturnicops noveboracensis; Yellow Rail; special concern, Alta, Man., NB, NS, NWT, Nunavut, Ont., Que., Sask.

Dendroica cerulea; Cerulean Warbler; special concern, Ont., Que.

Dendroica kirtlandii; Kirtland's Warbler; endangered, Ont., Que.

Ectopistes migratorius; Passenger Pigeon; extinct, Man., NB, NS, Ont., PEI, Que., Sask.

Empidonax virescens; Acadian Flycatcher; endangered, Ont.

Falco peregrinus anatum; Anatum Peregrine Falcon; threatened, Alta, BC, Man., NB, Nfld, NS, NWT, Nunavut, Ont., Que., Sask., Yukon

Falco peregrinus pealei; Peale's Peregrine Falcon; special concern, BC

Falco peregrinus tundrius; Tundra Peregrine Falcon; special concern, Nfld, NWT, Nunavut, Que., Yukon

Grus americana; Whooping Crane; endangered, NWT, Nunavut

Histrionicus histrionicus; Harlequin Duck; endangered, NB, Nfld, NS, Que. (eastern population)

Icteria virens auricollis; Yellow-Breasted Chat; threatened, BC population

Icteria virens virens; Yellow-Breasted Chat; special concern, Ont. (eastern population)

Ixobrychus exilis; Least Bittern; special concern, Man., NB, Ont., Que.

Lanius ludovicianus excubitorides; Loggerhead Shrike; threatened, Alta, Man., Sask. (western population)

Lanius ludovicianus migrans; Loggerhead Shrike; endangered, Man., Ont., Que. (eastern population)

Melanerpes erythrocephalus; Red-Headed Woodpecker; special concern, Man., Ont., Que., Sask.

Melanerpes lewis; Lewis's Woodpecker; special concern, BC

Numenius americanus; Long-Billed Curlew; special concern, Alta, BC, Sask.

Numenius borealis; Eskimo Curlew; endangered, all provinces and territories except BC

Oreoscoptes montanus; Sage Thrasher; endangered, Alta, BC, Sask.

Otus flammeolus; Flammulated Owl; special concern, BC

Pagophila eburnea; Ivory Gull; special concern; BC, Man., NB, Nfld, NS, NWT, Nunavut, Ont., Que.

Passerculus sandwichensis princeps; 'Ipswich' Savannah Sparrow; special concern, NS

Picoides albolarvatus; White-Headed Woodpecker; threatened, BC

Pinguinis impennis; Great Auk; extinct, NB, Nfld, NS, Que. (coastal waters)

Protonotaria citrea; Prothonotary Warbler; endangered, Ont.

Rallus elegans; King Rail, endangered, Ont.

Rhodostethia rosea; Ross's Gull; special concern, Man., NWT, Nunavut

Seiurus motacilla; Louisiana Waterthrush; special concern, Ont., Que.

Speotyto cunicularia; Burrowing Owl; endangered, Alta, BC, Man., Sask.
Sterna dougallii; Roseate Tern; endangered, NS, Que.
Strix occidentalis caurina; Northern Spotted Owl; endangered, BC
Synthliboramphus antiquus; Ancient Murrelet; special concern, BC
Tympanuchus cupido; Greater Prairie Chicken; extirpated, Alta, Man., Ont., Sask.
Tyto alba; Barn Owl; endangered, Ont. (eastern population)
Tyto alba; Barn Owl; special concern, BC (western population)
Wilsonia citrina; Hooded Warbler; threatened, Ont.

Mammals
Antrozous pallidus; Pallid Bat; threatened, BC
Balaena mysticetus; Bowhead Whale; endangered, Eastern Arctic
Balaena mysticetus; Bowhead Whale; endangered, Western Arctic
Balaenoptera musculus; Blue Whale; special concern, Atlantic and Pacific Oceans
Balaenoptera physalus; Fin Whale; special concern, Atlantic and Pacific Oceans
Bison bison athabascae; Wood Bison; threatened, Alta, BC, NWT, Yukon
Cynomys ludovicianus; Black-Tailed Prairie Dog; special concern, Sask.
Delphinapterus leucas; Beluga Whale; endangered, Que. (St Lawrence population)
Delphinapterus leucas; Beluga Whale; endangered, Nunavut (Ungava Bay)
Delphinapterus leucas; Beluga Whale; endangered, Nunavut (Cumberland Sound)
Delphinapterus leucas; Beluga Whale; threatened, Nunavut (eastern Hudson Bay)
Delphinapterus leucas; Beluga Whale; special concern, Nunavut (eastern High Arctic/Baffin Bay)
Dipodomys ordii; Ord's Kangaroo Rat; special concern, Alta, Sask.
Enhydra lutris; Sea Otter; threatened, BC (Pacific coast)
Eschrichtius robustus; Gray Whale; extirpated, Northwest Atlantic
Eubalaena glacialis; Right Whale; endangered, Atlantic and Pacific Oceans
Euderma maculatum; Spotted Bat; special concern, BC
Glaucomys volans; Southern Flying Squirrel; special concern, NB, NS, Ont., Que.
Gulo gulo; Wolverine; endangered, Que., Nfld (Labrador; eastern population)
Gulo gulo; Wolverine; special concern, Alta, BC, Man., NWT, Nunavut, Ont., Sask., Yukon (western population)
Hyperoodon ampullatus; Northern Bottlenose Whale; special concern, Gully population, Atlantic Ocean
Marmota vancouverensis; Vancouver Island Marmot; endangered, BC
Martes americana atrata; American Marten; endangered, Nfld (Nfld population)
Megaptera novaeangliae; Humpback Whale; threatened, North Pacific population
Megaptera novaeangliae; Humpback Whale; special concern, western North Atlantic population
Mesopledon bidens; Sowerby's Beaked Whale; special concern, Atlantic Ocean
Microtus pinetorum; Woodland Vole; special concern, Ont., Que.
Mustela erminea haidarum; Ermine; special concern, BC (Queen Charlotte Islands population)

Mustela macrodon; Sea Mink; extinct, NB, Nfld (coastal)

Mustela nigripes; Black-Footed Ferret; extirpated, Alta, Man., Sask.

Myotis keenii; Keen's Long-Eared Bat; special concern, BC

Myotis thysanodes; Fringed Myotis Bat; special concern, BC

Odobenus rosmarus rosmarus; Atlantic Walrus; extirpated, Northwest Atlantic

Orcinus orca; Killer Whale; threatened, North Pacific 'resident' populations

Orcinus orca; Killer Whale; special concern, North Pacific 'transient' populations

Phoca vitulina mellonae; Harbour Seal; special concern, Que. (Lac des Loups Marins population)

Phocoena phocoena; Harbour Porpoise; threatened, Northwest Atlantic population

Rangifer tarandus caribou; Woodland Caribou; endangered, Que. (Gaspé population)

Rangifer tarandus caribou; Woodland Caribou; threatened, Alta, BC, Man., Nfld, NWT, Ont., Que., Sask., Yukon (Boreal population)

Rangifer tarandus caribou; Woodland Caribou; threatened, Alta, BC (Southern Mountain population)

Rangifer tarandus caribou; Woodland Caribou; special concern, Alta, BC, Man., NWT, Nunavut, Ont., Sask. (western population)

Rangifer tarandus dawsoni; Queen Charlotte Woodland Caribou; extinct, BC

Rangifer tarandus pearyi; Peary Caribou; endangered, Nunavut (Banks Island)

Rangifer tarandus pearyi; Peary Caribou; endangered, Nunavut (High Arctic)

Rangifer tarandus pearyi; Peary Caribou; threatened, NWT, Nunavut (Low Arctic population)

Reithrodontomys megalotis megalotis; Western Harvest Mouse; special concern, BC (BC population)

Scalopus aquaticus; Eastern Mole; special concern, Ont.

Scapanus townsendii; Townsend's Mole; threatened, BC

Sorex bendirii; Pacific Water Shrew; threatened, BC

Sorex gaspensis; Gaspé Shrew; special concern, NB, NS, Que.

Sylvilagus nuttallii nuttallii; Nuttall's Cottontail; special concern; BC population

Taxidea taxus ssp. *jacksoni*; American Badger; endangered, Ont.

Taxidea taxus ssp. *jeffersonii*; American Badger; endangered, BC

Urocyon cinereoargenteus; Grey Fox; special concern, Alta, Man., Ont., Que.

Ursus arctos; Grizzly Bear; extirpated, Alta, Man., Sask. (Prairie population)

Ursus arctos; Grizzly Bear; special concern, Alta, BC, NWT, Nunavut, Yukon

Ursus maritimus; Polar Bear; special concern, Man., Nfld, NWT, Nunavut, Ont., Que.

Vulpes velox; Swift Fox; endangered, Alta, Sask.

CHAPTER THREE

Avoiding the Endangerment of Species: The Importance of Habitats and Ecosystems

DAVID GAUTHIER AND ED WIKEN

INTRODUCTION

The presence of endangered species in Canada represents a negative legacy. Resource management and planning activities over the last century have largely neglected the survival of many wild species and the importance of habitat. Endangered wildlife and habitats across much of the southern latitudes of Canada are the legacies of land and water development that has overly favoured resource exploitation and production. Dealing with the aftermath of habitat degradation and loss occupies much of the agenda concerning wildlife management activities today. More importantly, it also tests the directions of conservation ethics for tomorrow. In this chapter, we argue for the need to consider the basic causes behind the issues of why so many species have become endangered. The quality, quantity, and health of habitats are critical in preventing species from becoming endangered and, ultimately, for preserving the ecosystems that encompass both the species and their habitats. A focus solely on a species may misinterpret the array and interplay of habitat factors that affect the survival and sustainability of species.

There are numerous spatial and temporal scales for considering these arrays and interactions, and the most suitable scales for investigation will vary with the questions asked. Wildlife studies can vary from detailed site investigations conducted over relatively short time—months to a few years—to regional studies that range over years and perhaps decades. Within those spatial and time scales, the selection of the unit of study is important. Should it be the species alone, particular populations of species, communities of species, the habitats of species, or perhaps the ecosystems that embrace habitats? Also, in many areas of Canada where endangerment of wildlife and habitats is an issue, the impacts of human activities are critically important to understanding the causes of endangerment. Successful management of endangered habitats and species in such cases is ultimately a function of understanding and mitigating the impacts of human activities and, at times, adapting to them.

Our understanding of the ecology of habitats and species must be integrated with our understanding of the socio-economic, cultural, and political dimensions of human behaviour. The status of any species or habitat type, by itself, is a measure or indicator to be considered in the context of many other such indicators. Together, they can be used to assess the overall integrity and sustainability of ecosystems.

The Wealth of Wildlife

Canada enjoys a wealth of wildlife, from rattlesnakes to polar bears, from pronghorns to killer whales. Many species are year-round residents and others migrate great distances to places as far away as southern Argentina (Wiken and Gauthier, 1997). Canada hosts some of the world's largest numbers of far-ranging species, such as the northern caribou herds (over 3 million animals) that range across the northern territories from Yukon in the West through to Quebec and Labrador in the East. The diversity and numbers of species are a reflection of the country's wealth of habitats. Canada's landscape provides over 9.9 million square kilometers of different habitats and the seascapes provide another 5.4 million square kilometres (Wiken, 1999) within the 200-mile limit of Canadian jurisdiction. This vast area includes over 20 major habitat/ecosystem types and over 300 regional habitats/ecosystems. About 46 per cent of the land-based habitats are in temperate forested areas and 45 per cent are in Arctic settings. Human settlements and disturbance tend to be dominant in the southern latitudes of the country. Urban areas, however, occupy less than 1 per cent of the land surface and rural settled areas amount to about 15 per cent. The Arctic (53 per cent) and the Atlantic (39 per cent) Oceans are spatially dominant among Canada's seascapes. The area of Canadian jurisdiction along the Pacific coast is surprisingly small but rich in its habitat types.

The more imminent threats to endangered species are somewhat limited to smaller and southern portions of Canada, such as the Prairie and Mixed-wood ecozones. While restricted in size, these areas stand out as serious triggers or signals and, strikingly, impinge on some of the more productive and diverse habitats in the country. Thus, the problem of endangered species and degraded habitats has wide implications for all of Canada.

In such a well-educated and relatively wealthy nation as Canada, given its vast area and diversity of habitats, how could species endangerment become a prominent issue? Are endangered species merely the canaries in the mineshaft in relation to the overall dysfunction of ecosystems? Do species act as indicators of more fundamental habitat or ecosystem problems? Are jurisdictions and agencies in Canada really prepared to tackle the causes of endangerment of species, or are their policies and programs addressing solely its symptoms? Are jurisdictions (i.e., federal, provincial, First Nations, and territories) prepared to collaborate and co-ordinate management efforts to reduce endangerment?

The answers to these questions require perspectives and approaches that span traditional jurisdictional boundaries and short temporal scales. The concept of 'habitat' provides a useful means by which to address species concerns and the array of biophysical and human factors affecting those species.

Species and Their Homes

Organizations like Wildlife Habitat Canada (1999) operate on the premise that 'Without habitat . . . there is no wildlife. It's that simple.' Protecting species must start with the principle that the 'homes' or 'habitats' of wildlife are of fundamental importance. Habitats and the ecosystems from which they are derived are the life-sustaining systems for wild species. When the country's wildlife habitats undergo degradation, significant alterations, or finally disappear, the effects are mimicked in species.

As natural habitats are altered or replaced with other types, a different mix often replaces the old complement of species. When these consist of species regarded as 'nuisance', 'weed', or 'noxious' species, the change is typically deemed 'negative', as it is when unique habitats and species are lost. Having to deal with wildlife management issues when species are 'threatened' and 'endangered' is an indicator, by any standard, of the poor performance of management policies and actions. Indeed, when habitats and species become 'extinct', any efforts at management and governance have failed completely. Most of these situations in Canada have arisen from efforts that have historically been misguided or based on uninformed decision-making (Government of Canada, 1996). Potentially, wildlife habitat management and planning initiatives have better prospects now. For example, in the management of game species, the concept of landscape ecology has been long recognized for its importance to habitat and wildlife. However, early management efforts, as well as many current efforts, focused on single species. Game management policies and programs have evolved to encompass more inclusive approaches that focus on communities and habitats. In regard to endangerment of wildlife, however, species approaches are still dominant.

Endangered Species, Endangered Habitats

In a scientific and intuitive sense, 'endangered species' and 'endangered habitats' are practically synonymous. However, when species disappear and seriously dwindle in numbers, a sense of urgency outpaces anything of a similar nature for the sometimes subtle and incremental impacts that degrade the integrity of critical habitats. The public's attention is drawn to the 'homeless' and 'displaced' more than to the basic elements that provide a home for such species. For endangered species in the southern, settled areas of Canada, the transformation of landscapes has reached levels where there is little or no other habitat in which traditional species can sustain themselves.

While endangered species and endangered ecosystems have similar connotations, the terms differ in a strategic cause-to-effect relationship. The endangerment of species is primarily an indicator of endangered habitats. The clearing of landscapes, widespread removal of trees, draining of wetlands, damming of rivers, and alteration of wildlands have been accepted signs and measures of human progress. It is difficult not to associate a sense of geography to these places of progress. Images of the settled and southern regions of Canada readily emerge. In some cases, very specific locations such as the Carolinian forests and the Prairies come to mind. These areas have been prized in Canada's early phases of human settlement and resource development.

Typically, they have been the most productive landscapes in the country for enterprises such as agriculture and forestry, and are also the most desirable for human habitation. Strikingly, those aspects have not favoured the conservation of native species, habitats, or ecosystems.

PROTECTING SPECIES AND HABITATS

A Great Start: An Ending Gone Awry

While endangered species legislation is a relatively recent phenomenon in Canadian politics, the conservation of wildlife habitat and nature is not (Wiken et al., 1998). Legislative attempts to conserve critical and significant wildlife habitats and natural areas in Canada date to the 1870s (Gilbert and Dodds, 1987). An Act to protect bison was belatedly passed in 1877 but had little effect due to lack of enforcement. Laws to strengthen protection of endangered wood bison were passed in 1893. Last Mountain Lake Sanctuary in Saskatchewan, the first wildlife area in Canada and in North America, was created in 1887. Gilbert and Dodds (1987) trace the modern beginnings of federal wildlife management activities to the Commission of Conservation constituted under the Conservation Act (1909). In 1916, the Advisory Board on Wildlife Protection drafted the Migratory Bird Treaty, and in 1917 drafted the Migratory Bird Convention Act Regulations. Some of the first Canadian marine conservation areas were created as early as 1919 (Wiken et al., 1998). These conservation efforts early in the twentieth century seemed to bode well for Canada's future stewardship of its habitats, species, lands, and waters. Yet, at the beginning of the twenty-first century, Canadian efforts to protect species and their habitats appear to have fallen short.

From the perspective of other nations, Canada is often viewed for the most part to be a pristine and vast wilderness. Perceptions aside, however, evidence indicates that Canada's species and habitat diversity are often threatened (Government of Canada, 1996). In southern Canada, agricultural and urban activities have extensively transformed much of the former native grasslands and forests. Intensive land uses and management practices dominate those landscapes. While generally less intensively managed than agricultural ecosystems, much of the southern parts of Canada's deciduous and boreal forests have been subject to extensive and intensive forestry operations that in some cases have had significant impacts. While the pole-ward margin of the boreal forest ecosystems and many parts of the Canadian Arctic would, at first, fit the image of unspoiled landscapes, the long-distance transport of airborne pollutants (Figure 3.1) is known to be affecting those habitats (Wiken and Gauthier, 1997). Canada's oceans have also been subject to impacts from pollution and overharvesting. All of these changes, near and far, constitute a potential risk to species and habitats, and, indeed, to whole ecosystems.

Protected areas are considered to be the wilderness 'banks' of Canada and elsewhere. Their establishment was seen by many to be the assured way to protect nature, unique and typical habitats, and species. From the beginning, protected areas were thought to provide safe havens for wildlife. At the beginning of the century, protect-

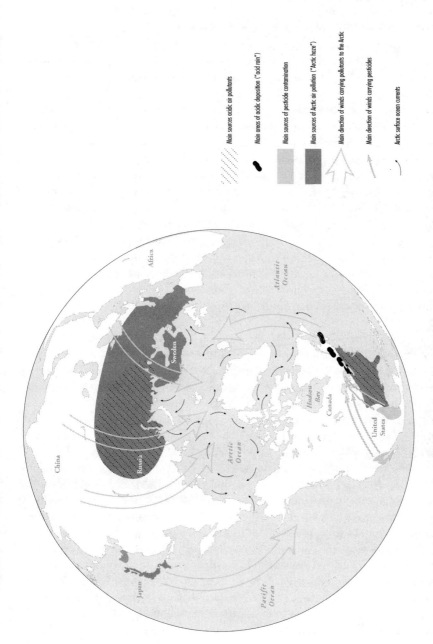

Figure 3.1: Distant Sources of Pollution Affecting Canadian Landscapes and Seascapes

ed areas were established in settings that were largely non-competitive with other sur-rounding land uses. The protected areas were not being heavily encroached by destructive land uses, laden with pollutants from distant sources, or permeated with roads and trails. In effect, they had many buffers that allowed them to maintain their ecological integrity. The park and wildlife offices set within the protected areas were in a way the 'frontier forts', trying to bring city-type order and rule to the wilderness. But like many of the frontier areas, they became engulfed by other types of foes—urbanization, forest operations, and farmland expansion. Currently, many of the southern protected areas have largely become isolated heritage areas, strangely dis-placed in time and purpose within a sea of human activities.

Presently, Canada has about 9 per cent (Government of Canada, 1996) of its landscapes in protected areas (Figure 3.2), much less is protected in the seascapes. Debate continues regarding what percentage is appropriate and what the minimal size of each protected area should be for maintaining core ecosystem functions and processes. Protected areas are too commonly seen as the only measure of con-servation. Instead, they should be viewed as one part of a spectrum of required conservation activities. For areas such as the Fraser Valley, the Prairies, and the St Lawrence Lowlands, managing and implementing conservation activities in non-protected areas are crucial. Initiatives such as private land stewardships and eco-logical gifting of private lands have become important avenues to enhance wildlife and habitats outside of formally protected areas. Planning for wildlife and habitats must be done within the context of the total landscape (or the greater ecosystem) and not just for areas called wildlife sanctuaries, parks, wilderness areas, and the like.

Human Settlements and Resource Developments Intervene

The impacts of humans and technologies have been similar to those of non-native or alien species. While humans as a biological species are a native part of many ecosys-tems, their technologies and activities have outstripped the capability of natural systems to accommodate the changes. Like zebra mussels and purple loosestrife, they spread by taking full advantage of their ability to subdue and displace other organ-isms, and consequently they divert the productivity of ecosystems to meet their needs and wants. The core ribbon of life for post-Aboriginal settlement in Canada became the thin but irregular band of territory along the 49th parallel (Brown, 1987). Human activity in that zone has engulfed productive ecosystems such as the grasslands of the Prairies, the Carolinian and mixed-wood forests of the St Lawrence Lowlands, the lower boreal forests, and the Fraser and Annapolis Valleys. Ironically, intensely used and developed areas are comparatively small (i.e., 10–20 per cent) in relation to Canada's total land mass. This small relative size, however, is confounded by two key factors: (1) within that southern ribbon, the diversity of species, ecosystems, and habi-tats in which human activity is concentrated is among the highest in the country; and (2) the ecosystems of which they are a part are among the rarest and most limited within the country.

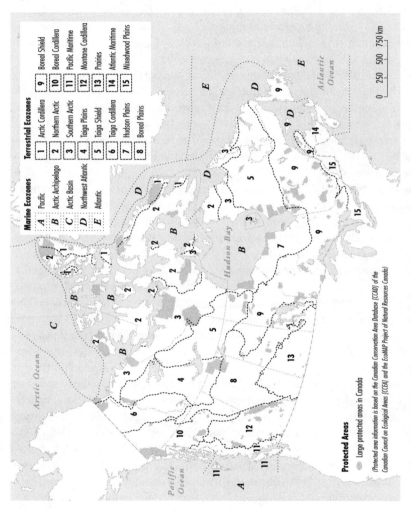

Figure 3.2: Marine and Terrestrial Ecozones, and Large Protected Areas of Canada

At the beginning of the twentieth century it must have been difficult to imagine that species or habitats in an area as vast as Canada would ever become endangered. Over the past 100 years, our perception of the vastness and apparent inexhaustibility of resources has fundamentally changed. The lessons learned include:

- the importance of critical connections at the continental and macro ecosystem levels—the notion that what was happening beyond a region, a province, and, at times, a nation could have critical impacts;
- the realization that pollution levels and impacts of land use could reach such inordinate levels of magnitude;
- the thresholds and breaking points of critical ecosystems and habitats could be breached under the pressure of human-generated stresses;
- that wildlife and habitats would reach unexpected degrees of economic and social importance;
- that seemingly minor or negligible by-products of human activities could accumulate in ecosystems and have significant negative long-term impacts on habitats and species;
- that to protect species, it is critical to protect their habitats and the ecosystems that contain those habitats.

What Exists to Protect Species and Their Habitats/Homes?
Legislation/Programs/Policies
Appendix A provides a summary of jurisdictional legislation, policies, and programs related to endangered species and habitats. Five provinces (Manitoba, Ontario, Quebec, New Brunswick, and Nova Scotia) have specific Acts directed towards endangered, threatened, or vulnerable species. A number of the provincial and territorial jurisdictions have similar Acts in development. All jurisdictions have wildlife and fisheries Acts, as well as Acts devoted to parks, natural areas, protected places, or ecological reserves. Saskatchewan is the only province to have a specific Wildlife Habitat Protection Act, although several jurisdictions have legislation to allow the designation of critical habitat (for example, Prince Edward Island and New Brunswick). Various sector-specific Acts are in place across the country, such as legislation specific to water or oceans, coastal habitats, forests, and wetlands. There is also more general legislation that incorporates ecosystem aspects, for example, respecting ecological resources (British Columbia) and heritage conservation (Manitoba, British Columbia, Ontario). Most jurisdictions have broad Environment Acts, and a few have specifically designated Environmental Assessment Acts (British Columbia, Saskatchewan, Ontario, Newfoundland).

On a positive note, the variety of legislation, policies, and programs signifies many means for protecting and conserving wildlife. However, it also means that there are numerous types and levels of authority. Appendix B provides a summary of gaps in legislation/policies/programs that was co-operatively prepared among the jurisdictions under the National Accord for the Protection of Species at Risk. That assessment indicates that at the provincial/territorial level there are some jurisdictions (for example, Prince Edward Island, Nova Scotia, and New Brunswick) that have enabled strong, mandatory legislation for species and habitat protection, while other jurisdictions have substantial gaps to fill. The majority of legislation related to protected areas is discretionary. Under provincial/territorial endangered species legislation, where such exists, the decision to list a species as endangered rests at the political level.

The federal government is the largest holder of conservation properties in Canada, with 30 million square kilometres. Much of the federal legislation designates that protected areas are to be managed by federal institutions and bureaucracies, such as the Canadian Wildlife Service, Parks Canada, and Fisheries and Oceans Canada (Lynch-Stewart et al., 1999). Many of these Acts were not initially intended to protect wildlife habitats (for example, the conservation of representative ecosystems may have been the primary goal) and those that did aim to protect habitats were limited largely to selective migratory species or to parts of the yearly life cycle of given species, as with Migratory Bird Sanctuaries (MBSS). One-third of Canada's federally protected lands (Beric, 1999) are administered under MBSS and National Wildlife Areas (NWAS) and are among the most poorly financed and monitored protected areas. Between these two designations, the MBSS account for most (90 per cent) of that property. Sanctuaries, by virtue of the name alone, conjure up images of safeguards for wildlife that would far exceed other types of protected areas. However, the reverse is true in that very little overall protection is provided to habitats under the MBS regulations. MBSS act as temporary spots with restricted safeguards for certain birds and their nests. The general habitat in these areas per se is not afforded much protection, nor is there long-term year-round protection.

While wildlife conservation is strongly supported by Canadians (Filion et al., 1995), the presumption is that conservation of species is automatic once protected areas have been set aside. Yet, while wildlife management and park agencies monitor species, for example, the monitoring is often selective and time-specific. Generally in Canada there is a lack of effective, comprehensive, ongoing monitoring of species or their habitats, or of comprehensive assessments of trends in changes in habitats. In particular, due consideration of the broader concerns of wildlife habitats before they reach critical points of species at risk is a major void.

DETERMINING HABITAT RISKS AND GAPS

An Ecosystem Basis

Canadians have been innovative and have achieved successes in assessing both human impacts and improved means of conservation management. The use of ecological frameworks as a basis for conservation management has contributed to those successes. When we say that without wildlife habitats there can be no wildlife, it is also appropriate to say that without ecosystems there can be no habitats. An ecosystem is considered to be an enduring system comprised of distinctive assemblages of organisms (including humans) and physical factors (soils, land forms, water, climate) (Wiken et al., 1996). Ecosystems occur at all scales and sizes, and exist as natural through to human-modified systems. They tend to change in a relatively consistent manner over time. Ecosystem and habitat management considers these principles and relationships.

In the fields of environmental studies and ecology, regions used to be seen as convenient categories for generalizing about the spatial and temporal occurrences of biological and physical characteristics. Increasingly, they are now seen as necessities in conceptualizing how other species and humans can successfully coexist within the world, both locally and globally. Fostering the development of a common ecosystem framework for marine and terrestrial habitats has been a key factor in helping to build practical and meaningful models for decision-making (Wiken and Gauthier, 1998a).

The ecoregion level of classification subdivides Canada into approximately 200 ecoregion units. These are further subdivided into approximately 1,000 ecodistricts. The ecoregions are also aggregated into 20 units called ecozones (Wiken et al., 1996). This ecological classification system for Canada constitutes the northern part of a North American ecological framework developed in co-operation with Mexico and the United States (NAEWG, 1997; Wiken and Gauthier, 1998b). This classification system provides an ecological framework within which to consider risks to biodiversity and habitats.

Assessment of Habitat Risks

Habitat risk assessments are useful in determining many threats but they also identify gaps. National risk studies, by design, are relative interpretations. This is not because they are abstract undertakings but rather because they must integrate various perspectives, interest groups, species, habitats, and ecosystems. Flexible approaches are needed in this regard to incorporate a variety of ideas and situations over time. A number of basic interpretive models can be developed based on core baseline data where it exists, but expert opinion must also be incorporated.

The process of creating a 'risk to wildlife habitats' and species evaluation includes the following elements: development of a conceptual framework; integration of relevant existing data sources and expertise; use of an integrated ecosystem approach; and development of a flexible approach that enables numerous scenarios to be run. Risk to habitats and species can be viewed as a function of: (1) the inherent characteristics of habitats/ecosystems; (2) the types of land and water uses that change habitats/ecosystems; (3) measures of the impacts of human activities; and (4) the success of conservation practices. Many of these factors were employed in Canada's earliest and most recent biodiversity risk assessments (Rubec et al., 1993; Turner et al., 1997).

Determining habitat risk can be immensely important to the conservation of habitats and species. Some of the potential uses are:

- as a tool to communicate the differences and concepts of habitat management and protection;
- as a foundation for the development of indicators to measure and track, over time, changes in the status of habitats;

- to help set national and regional priorities for focusing policies and programs that support the conservation and sustainable use of habitats; and
- to provide a catalyst for completing similar assessment at other scales, including North America.

Wildlife habitat, biodiversity, and ecosystem integrity assessments can provide valuable results for setting national and regional priorities. The concepts, however, are valuable at all scales. Figure 3.3 shows a national assessment of risks using the national ecoregion classification for the country as a framework. It reflects risks to wildlife habitats across the country and shows that the greatest immediate risks are in the more southern ecosystems and habitats of Canada. The high-risk areas are where there have been many land-management and planning gaps in properly caring for species and their habitats. It also shows areas where the threat remains high. Those areas tend to reflect management needs that fall within the 'react and cure' mode. The low-risk areas should not be considered as areas that can be ignored—they are simply areas that contain many wildlife assets that will remain in good condition provided that 'anticipate and prevent' modes of management are applied before habitat conditions reach the stages typical of southern Canada. Both modes of management depend on the quality of and access to wildlife information.

Wildlife Habitat Information
The information needs of wildlife managers and policy-makers have grown in complexity. However, the data required to support these information needs have either remained static or diminished as a result of the many cutbacks in wildlife organizations. Indeed, some monitoring and assessment programs have been abandoned. Most of the currently available data come from inventories and monitoring systems designed years ago to meet rather narrowly defined wildlife purposes, such as capability ratings and harvesting. Relatively little new comprehensive data that might support long-term resource concerns are being collected (Wiken and Gauthier, 1998a) and wildlife habitat issues are one of many examples. Risk assessments using ecosystem approaches are responding to a very different management paradigm from the traditional wildlife management issues of the past (Turner et al., 1997). Although volumes of habitat and species data exist, much of that data refers to questions and issues with little direct relevance to habitat/ecosystem-level questions and needs. A decade ago, for instance, the focus was very much on resource development and extraction. Inventories and monitoring systems were largely geared to those ends. The design of data systems was based on questions such as 'What harvestable species are there in landscapes and seascapes?' rather than 'What types of factors are required to sustain productive habitats and populations?'

Initiatives such as risk assessment must have access to newer and broader data sets capable of responding to the different management paradigm inherent in sustainable resource development. Integration of data is a key activity for the depiction and analysis of ecosystem events. There is no particular strength in isolated data, particu-

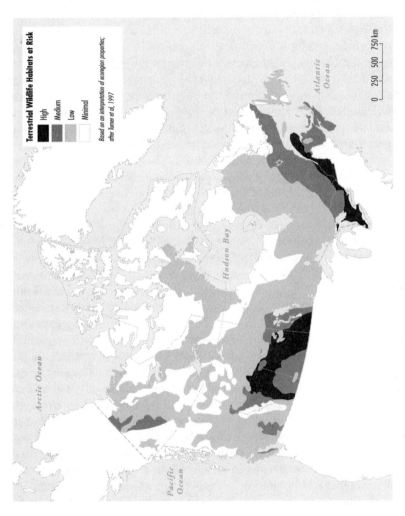

Figure 3.5: Terrestrial Wildlife Habitats at Risk, Canada

larly when an ecosystem perspective is required for so many applications. Data concerning human activities, land use, plant and animal species, ecosystems, climate, historical baselines, water bodies, and so on are often lacking, incomplete, and/or incompatible. Available habitat data often exist in different formats and scales requiring expensive conversions to be useful.

ECOSYSTEM MANAGEMENT FOR WILDLIFE

A Way Ahead

When acid rain became a policy issue in the 1970s, the discovery of dead fish in lakes was seen as symptomatic of a much larger problem. The resolution of the problem was not in replacing the fish but in understanding ecosystems and our role within them. The existence and growth in the number of endangered species are also symptomatic of larger problems. They are a signal and reflection of the increasing impacts of humans on ecosystems and the difficulties that humans have had in adjusting their actions to sustain the integrity of ecosystems. In recognition of these factors, the concerns of Canadians have evolved over time from an interest in parts of nature to questions of the integrity and health of nature as a whole. At the beginning of the twentieth century Canada could be characterized as insular areas of settlement surrounded by a wide buffer of wilderness; now, particularly in the southern parts of the country, the reverse is true: solitary protected areas represent the refuge of wilderness, so small and isolated that they typically have lost their capacity to support significant numbers of large animals.

Addressing the issue of endangered species is one way of drawing immediate attention to the recovery or prevention of further loss of species in places such as the southern and settled areas of Canada. Yet there is little practicality and sense over the long term of managing by individual species or even communities of species. An agonizing aspect of wildlife management is the need to recognize that there are insufficient human resources to manage all species. How do we determine which are the most deserving species? An estimated 282,000 species (see Table 2.2) are spread across the country's landscapes and seascapes and between provincial, international, territorial, and First Nation jurisdictions. Beyond the impracticability of dealing with such vast numbers, it is not sensible to manage isolated pieces. The way ahead must place a greater emphasis on integrated, comprehensive inventories and monitoring systems—ones specially designed to meet analytical needs at the scales of habitats and ecosystems. These data sets need to be ecosystem-based, scientifically sound, and collected over longer time frames. Today's conservation strategies and interests in endangered species and habitat issues have leapt ahead of the supporting data and information. It is a phenomenon that applies from regional and provincial/territorial levels through to the international scene.

Ecosystem management and science were once thought to be the forte of biologists and naturalists. Now they are the new tools and realities of decision-makers and resource managers. In that context, ecosystem management is best thought of as an approach that integrates our collective wisdom and actions. Instead of dividing up actions, jurisdictions, scientific approaches, and capabilities, it encourages the opposite to transpire. Ultimately, success will result from decision-making processes that follow such patterns. For example, wildlife managers and planners cannot effectively care for caribou without knowing their habitats and needs throughout the year, their life-cycle requirements, their tolerance to human activities and pollutants, their role in Native cultures, their importance to tourism and recreation, their contribution to biodiversity and ecosystem dynamics, and their relationship in the food chain. The connected parts and their relationships over temporal and spatial periods must all be understood.

POLITICS OF THE WILD

It is easy to understand why decision-makers would opt for goals that have immediate political payoffs but long-term habitat costs. Yet, the past century of our misunderstanding of the impacts of human activities in relation to the health of habitats and the ecosystems of which people are a part must be accounted for. Innovative and integrative solutions to resource management are required—approaches that put the highest priority on sustaining the integrity of habitats and ecosystems. Canada's wildlife habitats must be managed as resources and as functional areas that serve both nature and people. A reorganization of priorities will happen in one of two ways. Either wise people will recognize the need to integrate human demands and habitat goals within the context of the carrying capacity of Canada's ecosystems, or habitat degradation will further the overall degradation of ecosystems. Eventually, the costs of our human-induced impacts will exceed even the short-term benefits. People will be forced to react, albeit too late to regain much of what they have lost (Gauthier and Henry, 1989).

On the one hand, it is gratifying to see that endangered species considerations have moved to the forefront in discussions regarding the politics of the wild. On the other, however, society is faced with much broader issues of wildlife habitat health and integrity issues. Success in reducing the endangerment of wildlife will not be achieved without an integrated focus that at least:

- further combines partnerships and skills among governments, First Nations, industries, private landowners, and other land stewards;
- adopts hierarchical spatial and temporal approaches based on wildlife habitat and ecological principles;
- sets measurable objectives and goals, and provides a means to monitor them;
- integrates biophysical and socio-economic, cultural, and political considerations into habitat and resource management decision-making;
- operates according to principles of sustainable resource use, adaptive management, and ecosystem management.

Managing wildlife habitats sustainably is a type of contract made co-operatively among citizens and groups that commits them to meeting their own needs without seriously compromising both the rights and needs of others or the basic quality of the environment. Such a contract has three principal goals: to assure ecosystem integrity; to assure human health and well-being; and to assure natural resource conservation. Sustainability cannot be achieved without achieving all of those elements, nor can it be expected to succeed if the basic needs of humans are not met (Gauthier, 1999). Ecosystem management is a key approach to achieving habitat goals. It requires a shift in the focus of humans from the production of goods and services to sustaining the viability of systems that are necessary to deliver goods and services now and into the future (ESA, 1995). This approach, when applied to habitats, requires the commitment of all levels of government, businesses, industries, First Nations, and all citizens to think, plan, and act in terms of ecosystems. Approaches to resolving the endangerment of wildlife habitats and their associated species should therefore be seen as one component of an overall ecosystem management strategy aiming to achieve sustainability of resources.

Politics in Canada is often predicated on jurisdictional boundaries, such as regional municipalities, provinces, and nations. The concerns of wildlife management, whether in terms of species or habitats, often transcend boundaries, as is the case with such other issues as acid rain, climate change, and forest sustainability. Thus, politics of the wild is really about mobilizing jurisdictional authorities to operate on the basis of the ecosystems and habitats that they share.

The effort to establish a federal Species at Risk Act (SARA) is a helpful first step (Environment Canada, 1999). It does recognize that a narrow focus on 'endangered species' and 'federal lands' is insufficient, and calls for co-operative partnerships among land and water managers and for the protection of critical wildlife habitats. Yet, a species-by-species approach will not easily work—the real focus should shift to habitats/ecosystems at risk. 'Critical habitats' in SARA tends to mean the last of the least, but critical and important wildlife habitats should be assessed and managed throughout Canada before they reach vulnerable stages. Indeed, one must ask, what habitats are not 'critical' or important? The concerns about endangered species are about more than the dwindling number of organisms. Like the coal miner's canary, they signal our inability to manage and care for habitats and the ecosystems of which they are a part, our inability to manage and control the impacts of human activities, and our inability to plan for the long term. That Canadians appreciate wildlife has never really changed, but their attitude towards wildlife habitats must change.

REFERENCES

Beric, R. 1999. 'Overview of the Canadian Conservation Database (CCAD)', *CCEA Newsletter* (Ottawa) No. 12.

Brown, C., ed. 1987. *The Illustrated History of Canada*. Toronto: Lester and Orpen Dennys.

Canadian Council on Ecological Areas (CCEA) home page: http://www.ccea.org

Ecological Society of America (ESA). 1995. *The Report of the Ecological Society of America Committee on the Scientific Basis for Ecosystem Management*. Washington.

Environment Canada. 1999. 'Canada's Plan for Protecting Species at Risk: An Update'. Ottawa, http://www.ec.gc.ca

Filion, T., et al. 1995. *The Importance of Wildlife to Canadians*. Ottawa: Canadian Wildlife Service.

Gauthier, D.A., ed. 1992. *Framework for a Nation-wide System of Ecological Areas in Canada: Part 1 – A Strategy*. Canada Council on Ecological Areas Occasional Paper Series Number 12. Ottawa.

——. 1999. 'Sustainable Development, Ecotourism and Wildlife', in J.G. Nelson, R. Butler, and G. Wall, eds, *Tourism and Sustainable Development: A Civic Approach*, 2nd edn. Heritage Resources Centre Joint Publication Number 2. Waterloo, Ont.: Department of Geography, University of Waterloo, 113–34.

—— and J.D. Henry. 1989. 'Misunderstanding the Prairies', in M. Hummel, ed., *Endangered Spaces: The Future for Canada's Wilderness*. Toronto: Key Porter Books, 183–93.

——, K. Kavanagh, T. Beechey, L. Goulet, and E. Wiken. 1995. *Ecoregion Gap Analysis: Framework for Developing a Nation-wide System of Protected Ecological Areas*. Canadian Council on Ecological Areas Occasional Paper Series Number 13. Ottawa.

Gilbert, F.F., and D.G. Dodds. 1987. *The Philosophy and Practice of Wildlife Management*. Malabar, Fla: Krieger.

Government of Canada. 1996. 'Understanding Interdependencies', in *State of the Environment Report: Conserving Canada's Natural Legacy*. CD-ROM. Ottawa: Minister of Public Works and Government Services Canada.

Lynch-Stewart, Pauline, Ingrid Kessel-Taylor, and Clayton Rubec. 1999. *Wetlands and Government: Policy and Legislation for Wetland Conservation in Canada*. Issues Paper, No. 1. Ottawa: North American Wetlands Conservation Council.

North American Ecosystem Working Group (NAEWG). 1997. *Ecological Regions of North America*. Montreal: Commission for Environmental Cooperation.

Rubec, C.D.A., A.M. Turner, and E.B. Wiken. 1993. 'Integrated Modeling for Protected Areas and Biodiversity Assessment in Canada', in J. Marczyk and D. Johnson, eds, *Sustainable Landscapes*, Proceedings of the Third Symposium of the Canadian Society for Landscape Ecology and Management, *Polyscience*: 157–76.

Turner, T., E.B. Wiken, and H. Moore. 1997. In N. Munro and J. Martin Willison, eds, *Modelling Risk to Biodiversity in Canadian Ecosystem Approach*, Proceedings of the Third International Conference of the Science and Management of Protected Areas, Calgary: 12–16 May, 657–67.

Wiken, E.B. 1999. 'Casting the Bottom Line on the Blue Planet', in M. Gorman and J. Loo, eds, *Protected Areas and the Bottom Line*, Proceedings of the 1997 CCEA annual meeting, Fredericton.

—— and D. Gauthier. 1997. 'Conservation and Ecology in North America', in P. Jonker et al., eds, *Caring for Home Place: Protected Areas and Landscape Ecology*, Proceedings of the 1996 CCEA annual meeting. Regina: University Extension Press and the Canadian Plains Research Center, 5–15.

—— and ——. 1998a. 'Reporting on the State of Ecosystems: Experiences with Integrating Monitoring and State of the Environment Reporting Activities in Canada and North America', presentation to the North American Symposium, Towards a Unified Framework for the Inventorying and Monitoring of Forest Ecosystem Resources: Mexico/US Symposium, Guadalajara, Mexico.

—— and ——. 1998b. 'Ecological Regions of North America', in N.W.P. Munro and J.H.M. Willison, eds, *Linking Protected Areas with Working Landscapes Conserving Biodiversity*,

Proceedings of the Third International Conference on Science and Management of Protected Areas, 12–16 May 1997. Wolfville, NS, 114–29.

——, ——, I. Marshall, K. Lawton, and H. Hirvonen. 1996. *A Perspective on Canada's Ecosystems: An Overview of the Terrestrial and Marine Ecozones.* Canadian Council on Ecological Areas Occasional Paper Number 14. Ottawa.

——, J. Robinson, and L. Warren. 1998. 'Return to the Sea: Conservation of Canadian Marine and Freshwater Ecosystems for Wildlife', paper presented at Marine Heritage Conservation Areas Workshop, 3 Apr. 1998, University of Waterloo.

Wildlife Habitat Canada. 1999. 'Wildlife Habitat Canada 15th Anniversary'. Ottawa (pamphlet).

APPENDIX A: LEGISLATION, POLICY, AND PROGRAMS BY JURISDICTION

Federal Government

Accord for the Protection of Species at Risk in Canada, 1998
Canada Oceans Act
Canada Wildlife Act
Canadian Environmental Assessment Act
Canadian Heritage Act
Canadian Wetland Classification System
Committee on the Status of Endangered Wildlife in Canada (COSEWIC)
Endangered Species Recovery Fund
Federal Policy on Wetland Conservation, 1991
Fisheries Act
Income Tax Act of Canada, 1995 (Ecological Gifts Program)
Marine Conservation Areas Act
Migratory Birds Convention Act
National Parks Act
North American Wetlands Conservation Council (Canada) report 'Canadian Legislation for Conservation Covenants, Easements and Servitude: The Current Situation, 1995'
Policy for the Management of Fish Habitat, 1986
Protecting Species at Risk in the NWT (pending)
Recovery of Nationally Endangered Wildlife (RENEW)
Wetland Policy
Wild Animal and Plant Protection, 1992
Wildlife Policy for Canada

Yukon

Accord for the Protection of Species at Risk in Canada, 1998
COSEWIC
Environmental Act
Federal Policy on Wetland Conservation, 1991
Forest Protection Act
Protected Areas Strategy (YPAS), 1998

RENEW
Territorial Parks Act
Yukon Endangered Species Protection Legislation (pending)
Yukon Wildlife Act, 1998 (amendments: Habitat Protection Areas (HPA), and
   Specially Protected Wildlife)

Northwest Territories
(Most of the NWT legislation applies to Nunavut.)
Accord for the Protection of Species at Risk in Canada, 1998
Federal Policy on Wetland Conservation, 1991
Forest Management Act
NWT Species at Risk Act (draft)
Planning Act
Territorial Parks Act
Wildlife Act
Wildlife Management Regulations: Outfitters, Muskox, Polar Bear, Barren Ground
   Caribou, Grizzly Bear, Wood Bison
Wildlife Regulations: Sanctuaries, Preserves, Regions

British Columbia
Legislation
Ecological Reserves Act
Ecological Resources Act
Environmental Assessment Act
Fish Protection Act
Forest Practices Code of British Columbia Act
Heritage Conservation Act
Land Act
Land Title Act
Municipal Act
Parks Act
Water Act
Waste Management Act
Wildlife Act

Policies/Programs
Accord for the Protection of Species at Risk in Canada, 1998 (participant)
Canadian Endangered Species Conservation Council (CESCC) (participant)
Committee on the Status of Endangered Wildlife (COSEWIC) (participant)
Identified Wildlife Management Strategy (Forest Practices Code)
Land and Resource Management Planning
Marbled Murrelet National Recovery Plan
Provincial Grizzly Bear Conservation Strategy

Multi-jurisdictional Co-operation for the Protection of Species that Cross Borders
Interagency Grizzly Bear Committee, Grizzly Initiative (with Parks Canada)
Anatum Peregrine Falcon, Marbled Murrelet, Wood Bison, south Okanagan Ecosystem Recovery Planning and Mountain Caribou Strategy
Recovery Teams for Garry Oak Ecosystems, Oregon Spotted Frog, and the Hotwater Physa

Alberta
Legislation
Environmental Protection and Enhancement Act, 1993
Federal Fisheries Act
Federal Migratory Birds Convention Act
Forests Reserves Act
Municipal Government Act (pending)
Provincial Parks Act
Public Lands Act
Water Act
Wild Animal and Plant Protection and Regulation of International and Inter-provincial Trade Act
Wilderness Areas, Ecological Reserves and Natural Areas Act
Wildlife Act (Alberta Endangered Species Conservation Committee, 1998)
Willmore Wilderness Park Act

Conservation Programs/Information/Research/Initiatives
Adjusting Grassland Insect Control Practices to Allow Survival of Grassland Songbirds
Alberta Environmentally Sustainable Agriculture Program
Alberta Forest Conservation Strategy
Alberta Riparian Habitat Management Project, Fisheries and Wildlife Management
    Canadian Biodiversity Strategy (participant)
Alberta Natural Heritage Information Centre
Biological Species/Observation Database
Buck for Wildlife Program
Canada/Alberta Environmentally Sustainable Agriculture Agreement's Constructed Wetlands for Treatment of Agricultural Runoff
Clean Air Strategic Alliance
Commercial Fishing Licensing Policy and the Fish Conservation Strategy for Alberta
Endangered Species Recovery Program
Fish Conservation Strategy for Alberta
Fisheries Management Enhancement Program
Foothills Model Forest
Instream Flow Needs Studies
Inventory of Private Land Forest Resource and Development of the Farm Woodlot Initiative

North America Waterfowl Management Plan
Prairie Conservation Action Plan
Public Lands Reservation/Notations Program
Purple Loosestrife Management Program
Rangeland Benchmark Programs (Benchmark Program and Rangeland Reference Area Program)
Regional Caribou Standing Committees
Retention of Wetlands and Wetland Habitat
Southern Rockies Landscape Planning Pilot Project,
Status of Wildlife (reports)
Sustainable Forest Management Network of Centres of Excellence

Policies
A Policy for Resource Management of the Eastern Slopes
Alberta Forest Conservation Strategy
Alberta Forest Legacy
Alberta Prairie Conservation Action Plan
Fish and Wildlife Policy for Alberta
Green Area and White Area Policy
Northern River Basins Study
Peace Athabasca Delta Technical Studies
Recommended Wetland Policy for Alberta
South Saskatchewan River Basin Policy
Special Places
Timber Harvest Planning and Operating Ground Rules
Wetland Management in the Settled Area of Alberta: An Interim Policy, 1993

Saskatchewan
Accord for the Protection of Species at Risk in Canada, 1998
Biodiversity Action Plan (in draft)
Conservation Easements Act
Ecological Reserves Act
Environmental Assessment Act
Environmental Management and Enhancement Act
Fisheries Act
Natural Resources Act
Parks Act
Planning and Development Act
Prairie Conservation Action Plan
Provincial Land Act
Public Lands Act
The Water Corporation Act
The Wild Species at Risk Regulations (Wildlife Act)

Wetland Policy, 1995
Wildlife Act, 1997
Wildlife Habitat Protection Act, 1992

Manitoba
Accord for the Protection of Species at Risk in Canada, 1998
Conservation Agreements Act (pending)
Ecological Reserves Act
Endangered Species Act
Environment Act
Forest Act
Heritage Resources Act
Manitoba Habitat Heritage Act
Manitoba Water Rights Act
Manitoba Water Policies, 1990
Municipal Act
Planning Act
Prairie Conservation Action Plan
Provincial Parks and Lands Act
Water Resources Administration Act
Wildlife Act

Ontario
Canadian Biodiversity Strategy
Conservation Land Act
Crown Forest Sustainability Act, 1998
Endangered Species Act, 1971
Environmental Assessment Act
Environmental Protection Act
Featured Species Policy (identifies endangered and threatened species as 'provincially
    featured species' in forest management planning)
Fish and Wildlife Conservation Act
Lakes and Rivers Improvement Act
Ministry of Government Services Act
Municipal Act
Natural Heritage Policies (provincial policy statement)
Ontario Heritage Act
Ontario Water Resources Act
Ontario Wetland Evaluation System, 1993
Planning Act
Provincial Parks Act
Provincial Policy Statement, 1997
Public Lands Act

Quebec
An Act Respecting the Conservation and Development of Wildlife
An Act Respecting Threatened or Vulnerable Species
Biodiversity Strategy and Action Plan, 1996
Canadian Biodiversity Strategy (member)
Ecological Reserves Act
Environmental Quality Act
Forest Act
Parks Act
Plant Protection Act

New Brunswick
Accord for the Protection of Species at Risk in Canada, 1998 (participant)
Clean Environment Act
Clean Water Act (Water Course Alteration Regulation)
Community Planning Act
Conservation Easements Act
Crown Lands and Forests Act
Ecological Reserves Act
Endangered Species Act, 1996
Fish and Wildlife Act, 1980
Forest Habitat Program
Land Use Policy for Coastal Lands (under Community Planning Act)
Parks Act
Protected Areas Program
Provincial Policy on Wetlands (draft)
Species at Risk Program
Wetlands and Coastal Habitat Program

Nova Scotia
Activities Designation Regulations
Conservation Easements Act
Crown Lands Act
Endangered Species Act, 1998
Environment Act
Environmental Assessment Regulations (under the Environment Act)
Planning Act
Provincial Parks Act
Special Places Protection Act
Water Act
Wetland Directive, 1995
Wildlife Act

Prince Edward Island
Accord for the Protection of Species at Risk in Canada, 1998
An Act to Amend the Natural Area Protection Act, 1992
Environmental Protection Act
Fish and Game Protection Act
Forest Management Act
Natural Area Protection Act, 1988
Planning Act
Significant Environmental Areas Program
Watercourse and Wetland Alteration Guidelines, 1995
Wildlife Conservation Act, 1998

Newfoundland and Labrador
Crown Lands Act
Department of Environment and Lands Act (Water Resources Policy)
Environmental Assessment Act
Environment and Lands Act
Forestry Act
Municipalities Act (together with the Eastern Habitat Joint Venture of North
    American Waterfowl Management Plan)
Provincial Parks Act
Waters Protection Act
Wilderness and Ecological Reserves Act
Wildlife Act

Examples of Private/Industry Sector Policies
Whitehorse Mining Initiative (Canadian Mining Association)
Preservation and Reclamation Policy (Canadian Sphagnum Peat Moss Association,
    1991)
Wetland Policy Statement (Canadian Pulp and Paper Association, 1992)

APPENDIX B: SUMMARY OF 'GAP ANALYSIS REPORT ON THE ACCORD FOR THE PROTECTION OF SPECIES AT RISK'

Yukon Territory
Situation: Endangered species legislation and land claims development assessment
process are currently being developed.
Gaps: There are many acknowledged gaps, and some relate to jurisdictional issues.
Without specific proposals being available, it is largely unclear whether new measures
will be mandatory or discretionary.
Remedies: Most gaps are expected to be adequately remedied by passing stand-alone
endangered species legislation, although the details of proposals are not yet available.
Verify: Conformity of new legislation with Accord; public relations strategy; encour-

agement of citizen involvement; resolution of jurisdictional issues; and extent of conservation activity, habitat protection, and enforcement.

Northwest Territories
Situation: Endangered species legislation is currently being developed, while settling and implementing land claims and devolution of federal powers are in process.
Gaps: Most gaps are acknowledged and are to be adequately addressed, and some relate to jurisdictional issues. Without specific proposals being available, it is largely unclear whether new measures will be mandatory or discretionary.
Remedies: Largely adequate remedies are proposed for the new legislation.
Verify: Passage of legislation, development of programs, and implementation.

British Columbia
Situation: Existing and newly expanded wildlife legislation is being augmented, and new forestry and fish habitat initiatives are getting under way.
Gaps: There are some remaining partial gaps and limiting factors affecting the scope of measures. Many key provisions are discretionary.
Remedies: Some new remedies are being developed, but further measures are needed to fully address identified gaps.
Verify: Listing of species, designation of habitat, and implementation of new initiatives.

Alberta
Situation: Generally well-developed legislation and programs in place.
Gaps: There are few gaps, with some gaps or partial gaps unacknowledged. Key provisions are mostly discretionary, except establishment of a scientific committee.
Remedies: Of the few required, partial or incomplete remedies were proposed.
Verify: Implementation of programs.

Saskatchewan
Situation: New Wildlife Act, 1997 and Wild Species at Risk Regulation incorporate almost all of the Accord provisions.
Gaps: Some partial gaps exist, and most are fully or partly acknowledged. Most key provisions are discretionary.
Remedies: Most gaps have proposed, sometimes partial, remedies.
Verify: Habitat protection; recovery plan process; implementation of monitoring and reporting process; preventive measures for vulnerable species; awareness/education strategy; extent of conservation activity, habitat protection, and enforcement.

Manitoba
Situation: Endangered Species Act passed in 1990. Legislative amendments will be proposed following completion of the Sustainable Development Initiative.
Gaps: There are several partial gaps, with most being acknowledged. Key provisions are half discretionary, half mandatory.

Remedies: Most remedies require further development.

Verify: Critical habitat designated; development and timely implementation of recovery plans; preventive measures for vulnerable species; communication programs; environmental assessment provisions; extent of conservation activity, habitat protection, and enforcement.

Ontario

Situation: The Fish and Wildlife Conservation Act and some implementing regulations recently came into force. Long-standing Endangered Species Act and species at risk programs are in place.

Gaps: Many gaps are partially acknowledged after previous study, yet some still remain due in part to need to better correlate legislation. Many key provisions are discretionary.

Remedies: Many remedies are partial and could be improved, while some gaps are identified but no or unclear remedies are proposed.

Verify: Endangered Species Act amendments, implementation of municipal planning and tax incentive programs, establishment of protected areas and habitat through the Lands for Life/Living Legacy process.

Quebec

Situation: The province is updating existing programs and revising existing legislation.

Gaps: Most gaps are acknowledged, although several only partly so. Most key provisions are discretionary.

Remedies: Some identified remedies need further development.

Verify: Refined definitions, private stewardship strategy, ecological servitudes, Quebec Conservation Data Centre, and publication of a new list of species.

New Brunswick

Situation: Recent updating of legislation and adjustments to meet the Accord.

Gaps: A few gaps remain, with most acknowledged or largely so. Many key provisions are mandatory.

Remedies: Fully or mostly adequate remedies are presented.

Verify: Reforms proposed, committee review and operation, the application of the Fish and Wildlife Act to threatened and vulnerable species, implementation and scope of status monitoring, communications/awareness strategy.

Nova Scotia

Situation: A strong Bill 65 was recently passed that corresponds closely to the Accord.

Gaps: Few gaps remain after new legislation. Most key provisions are mandatory.

Remedies: The few required remedies are largely adequate.

Verify: Implementation of the new Endangered Species Act and development of related regulations (including those for protecting critical habitat) and programs.

Prince Edward Island
Situation: Recent passage of Wildlife Conservation Act, with plans to develop and implement programs through leading charitable land trust.
Gaps: Some unacknowledged full and partial gaps are noted, including technical interpretation of legislation. Key provisions are mostly discretionary.
Remedies: Of the remedies required, most are partial and need further development.
Verify: Regulations, committee, program and private delivery partnership, habitat determinations, and environmental assessment review.

Newfoundland and Labrador
Situation: After completion of consultation on legislative proposal, the province is now preparing a draft Act and anticipating development of further programs.
Gaps: Many gaps are present, and they are only partly acknowledged. Key proposed provisions are split between mandatory and discretionary.
Remedies: Many gaps are to be addressed adequately by new legislation, but some remedies can be improved or clarified.
Verify: Passage of Act and regulations, program development.

Endangered Species and Terrestrial Protected Areas

PHILIP DEARDEN

INTRODUCTION

Protected areas in general are set aside because the values they contain would not survive were they not protected from the market forces that rule decision-making and resource allocation on the rest of the landscape. The nature of these values is broad, ranging from recreational, spiritual, and aesthetic dimensions to protection of ecosystem processes and components. Over the past decade the relative balance between these different values has shifted somewhat, with greater attention, certainly at the national and provincial levels, now being given to environmental conservation values rather than to direct use values, such as recreation (McNamee, 1993). One of the main reasons for this is the scientific consensus regarding the serious nature of continued environmental degradation. Symptomatic of this degradation are the increasing numbers of species under threat of extinction, both globally and in Canada. Habitat destruction is, overall, the main cause of endangerment (Foin et al., 1998). Protected areas have thus emerged as a key factor in global efforts to stem rising extinction rates.

This chapter provides an overview of the role of terrestrial protected areas in endangered species protection in Canada, reviews some of the main threats being faced by protected areas, and offers some thoughts for future challenges.

PROTECTED AREAS IN CANADA

National Parks

Legislation

There is no explicit reference to endangered species in the National Parks Act. However, insofar as the dedication clause calls for the parks to be left 'unimpaired for the enjoyment of future generations', it is apparent that any loss of species would violate the Act. Since, by definition, endangered species are those most likely to succumb, then an argument could be made that there is a legislative mandate for special attention to be paid to endangered species. Such a legislative mandate would be further strengthened by the 1988 amendments to the Act (s. 512), which state that 'Maintenance of ecological integrity through the protection of natural resources shall be the first priority when considering Park zoning and visitor use in a management

Table 4.1: National Parks of Canada

Name	Province	Established	Size (km²)
Aulavik	NWT	1992	12,200
Tuktut	NWT	1996	16,340
Nahanni	NWT	1972	4,766
Wood Buffalo	NWT	1922	44,807
Quttinirpaaq	Nunavut	1998	37,775
Sirmilik	Nunavut	1999	21,000
Auyuittuq	Nunavut	1976	19,700
Ivvavik	Yukon	1984	10,170
Kluane	Yukon	1972	22,015
Gwaii Haanas	BC	1993	1,470
Pacific Rim	BC	2000	499
Mt Revelstoke	BC	1914	260
Glacier	BC	1886	1,349
Yoho	BC	1896	1,310
Kootenay	BC	1920	1,406
Jasper	Alberta	1907	10,878
Banff	Alberta	1885	6,641
Elk Island	Alberta	1913	194
Waterton	Alberta	1895	525
Prince Albert	Sask	1927	3,875
Grasslands	Sask	1988	450*
Riding Mountain	Manitoba	1930	2,973
Wapusk	Manitoba	1996	11,475
Pukaskwa	Ontario	1983	1,878
Bruce Peninsula	Ontario	2000	136
Georgian Bay	Ontario	1929	14
Point Pelee	Ontario	1918	20
St Lawrence Islands	Ontario	1914	9
Mingan Archipelago	Quebec	1984	150
Forillon	Quebec	1970	244
La Maurice	Quebec	1970	536
Kouchibouguac	New Brunswick	1969	238
Fundy	New Brunswick	1948	206
Prince Edward Island	Prince Edward Island	1937	18
Cape Breton Highlands	Nova Scotia	1936	950
Kejimkujik	Nova Scotia	1969	403
Gros Morne	Newfoundland	1973	1,805
Terra Nova	Newfoundland	1957	404

*Still acquiring private lands, will grow to 900 km².

plan'. This ecological emphasis was further strengthened in the 1998 Parks Canada Agency Act. Thus, although there may not be explicit reference to endangered species in the legislation, it would be difficult to envisage the legislation not implicitly including the protection of endangered species as a significant part of the mandate.

Policy

The location of Canada's national parks (see Table 4.1) is in accordance with a systems plan that seeks to represent each of the 39 natural regions in the country. In selecting which areas within each natural region might be considered as park candidates, it is required that 'the occurrence of rare, threatened or endangered wildlife and vegetation' be taken into account (Parks Canada, 1994: 1.2.2). How critical such factors are in the final choice of site and boundaries varies from park to park. Some parks had their origins firmly linked with the need to protect specific species. Wood Buffalo National Park, for example, was first created in 1922 to protect the dwindling numbers of wood buffalo and within this vast park was later discovered not only a genetically purer strain of wood buffalo, but also the nesting site for the highly endangered whooping crane.

In more recent times, the boundaries for the proposed new national park in the Torngat Mountains in northern Labrador have been influenced by the need to take into account the breeding ranges of species at risk, such as the harlequin duck, peregrine falcon, and barren ground caribou. However, relative to economic and political dictates, the boundaries of the vast majority of parks were established with little attention to species requirements. In Nahanni National Park Reserve in the Northwest Territories, for example, although caribou winter in the park, their critical calving grounds are to the northwest on the Yukon border. The new park now being established in the Interlake region in Manitoba suffers from the same deficiency, with the most important woodland caribou habitat excluded.

In addition to the systems plan, Parks Canada legislation and policy clearly state that the protection of ecological integrity is the primary function of the national parks. For example, the policy document states that: 'Protecting ecological integrity and ensuring commemorative integrity take priority in acquiring, managing and administering heritage places and programs. In every application of policy, this guiding principle is paramount' (ibid.). If such is the case then it would appear that Parks Canada has a very strong mandate in both legislation and policy for special attention to be directed to the protection of endangered species. In reality, however, several studies (Wipond and Dearden, 1996; Auditor General, 1996; Minister of Public Works and Government Services Canada, 1998; Parks Canada Agency, 2000) have pointed out that ecological integrity is severely compromised in many parks.

Management

Management Directive 2.4.3 of Parks Canada states that 'Parks lands will be managed to assure the protection of indigenous and reintroduced rare, threatened and endangered species' (Parks Canada, 1986: 6). One of the main means of achieving this is through the internal zoning system for parks. In the five-zone system the first two zones are of special relevance to species at risk. Zone 1 is the 'Special Preservation' zone, which may contain or support unique, threatened, or endangered species, and in which motorized access is not permitted and public access may be limited. Such

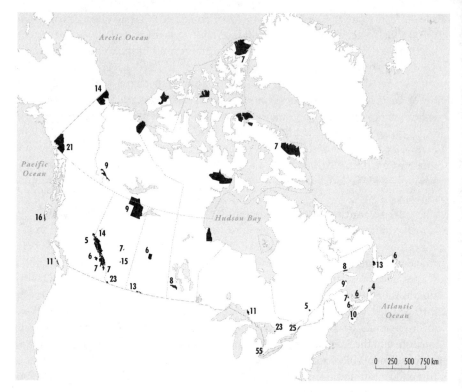

Figure 4.1: Number of COSEWIC Species in National Parks

limitations are in effect, for example, for the whooping crane nesting sites in Wood Buffalo, for the white pelican in Prince Albert National Park, and for the piping plover in Prince Edward Island National Park. In Zone 2, the 'Wilderness' zone, there is to be minimal human interference, with no motorized access and few services. Together, these two zones should comprise the majority of the park (Parks Canada, 1994). In Banff, for example, 4 per cent of the total park area is in Zone 1 and 93 per cent in Zone 2. For Wood Buffalo, the respective figures are 10 and 86 per cent.

However, the zoning system is not without fault. In many cases zoning is based on past uses rather than ecological value, is quite coarse, and may miss small but significant resources. In Banff this was somewhat addressed in the management plan by designation of a new zone, 'Environmentally Sensitive Sites', which includes such areas as the Vermillion Lakes Wetlands and the Middle Springs on Sulphur Mountain. Furthermore, although the National Parks Act has a provision under section 14 for the creation of wilderness zones by regulation, thereby preventing any inappropriate development or activities in the zone, as of yet no such designation has been enacted.

Table 4.2: The Role of National Parks in the Protection of Designated Vascular and Vertebrate Species at Risk in Canada

	Native Vascular Plant Species	Vascular Plant Species at Risk*	Native Vertebrate Species	Vertebrate Species at Risk*	Square Kilometres
Canada	4,521	109	1,061	190	9,900,000
National Parks	3,192	62	858	92	256,385
% of Canada total in Parks	70.6	56.9	80.9	48.4	2.59

*Species at risk in Canada (of special concern, threatened, endangered) as designated by the Committee on the Status of Endangered Wildlife in Canada (COSEWIC).
SOURCE: Parks Canada, personal communication, 1999, based on data available in 1998.

Besides zoning, specific management actions for wildlife in general and for certain key species are also specified in individual park management plans. In Banff, for example, increased attention is now being paid to the impacts of park development on the activities of predators such as wolves and grizzly bears. Of particular concern with respect to wildlife movement is the Cascade Corridor between Cascade Mountain and the Trans-Canada Highway. The Timberline Lodge, the road to Mount Norquay, Forty Mile Creek Reservoir, horse corrals, a bison paddock, an airstrip, a road to Lake Minnewanka, and an Army cadet camp are all located here. As a result the Banff Management Plan proposed to close the airstrip, relocate the horse corrals and cadet camp, remove the bison paddock, ensure that the Timberline Lodge does not expand beyond its current developed footprint, and, in co-operation with the operators of Mount Norquay, monitor wildlife movement in the area and use information from this study to determine if changes to summer use of the Norquay access road are needed. Except for closing the airstrip, which has become a legal question, all of these activities have been undertaken.

Role in Endangered Species Protection
Overall, Canada's national park system plays a crucial role in the protection of rare and endangered species (Figure 4.1). Although the 39 parks cover only 2.6 per cent of the Canadian land mass they contain 70.6 per cent of the native terrestrial and freshwater vascular plant species and 80.9 per cent of the vertebrate species (Table 4.2). This high degree of coverage is a result of the explicit systems plan of Parks Canada to establish parks in the 39 representative natural regions of the country. Almost 50 per cent of the endangered vertebrate species can be found in parks, as are 57 per cent of vascular plant species (Table 4.2). The national parks have also played a critical and increasingly important role as sites for the reintroduction of species (Table 4.3).

Wood Buffalo is a good example of a park that has played an important role in endangered species protection and recovery. At 44,807 square kilometres, it is the sec-

Table 4.3: Examples of Reintroductions of Species into Canadian National Parks

Species	Park
American Beaver	Cape Breton Highlands National Park; Prince Edward Island National Park
American Bison	Prince Albert National Park; Riding Mountain National Park
Plains Bison	Elk Island National Park
Wood Bison	Nahanni National Park Reserve; Jasper National Park; Waterton Lakes National Park; Elk Island National Park
Fisher	Georgian Bay Islands National Park; Riding Mountain National Park; Elk Island National Park
American Marten	Fundy National Park; Kejimkujik National Park; Terra Nova National Park; Riding Mountain National Park
Moose	Cape Breton Highlands National Park
Muskox	Ivvavik National Park
Trumpeter Swan	Elk Island National Park
Caribou	Cape Breton Highlands National Park
Swift Fox	Grassland National Park

ond largest terrestrial national park in the world (the largest is in Greenland). It was designated in 1922 largely to protect the remaining wood bison to be found there. It later became the home for a transplanted herd of Plains bison bought by the Canadian government from an American rancher, following the extirpation of the species in Canada. These two species interbred, and it was feared that although the total number of bison had grown significantly, the distinctive characteristics of the wood bison had been lost. In 1957, however, an isolated group of wood bison was located in a remote area of this vast park. This herd was relocated to two other locations to guard against further interbreeding. The herd has continued to expand and has provided important satellite populations for other parks, such as Elk Island National Park outside Edmonton, and also in Manitoba, Yukon, and the Northwest Territories. In 1988 the status of the wood bison was changed from endangered to threatened by COSEWIC.

Another endangered species to benefit from the protection offered by Wood Buffalo is the whooping crane. Unlike the bison, they were never very numerous, and as their main nesting grounds on prairie marshes became increasingly disturbed their numbers dwindled rapidly. Unrestricted hunting along their long migration routes from the Prairies to the Gulf of Mexico also contributed to their near-extinction. By 1941, only 22 whooping cranes remained, and government action in both the US and Canada was started to try to save the species from extinction. The 1916 Migratory Bird Treaty between the countries was used to stop legal hunting and in 1937 the US government established the Aransas National Wildlife Refuge to protect wintering habitat along the Gulf Coast. The big breakthrough, however, occurred in 1954 when the

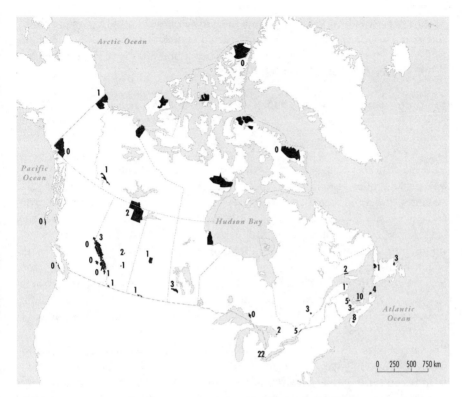

Figure 4.2: Number of Vertebrate Animal Species Extirpated from National Parks
SOURCE: Minister of Public Works and Government Services Canada (1998).

only known nesting area was discovered in the northern part of Wood Buffalo. This allowed not only greater protection for their breeding grounds but also active manipulation, such as artificial egg incubation. As a result of these measures the total population has risen to over 300, with a second breeding population now established in Manitoba.

National parks are not always successful in protecting populations of species at risk, however (Figure 4.2). Extirpations from parks include species such as wolf, beaver, bison, cougar, lynx, marten, and caribou, to name but a few of the vertebrates. Extinctions have also occurred. The Banff longnose dace was a small minnow found in only one location on earth, a marsh downstream from the famous Cave and Basin hot spring in what is now Banff National Park. Discovered in 1892, it was declared extinct by COSEWIC in 1987. As with many extinctions, a combination of factors likely led to its demise. Tropical fish were introduced into the warm waters of the marsh, where they flourished and were able to out-compete the dace. The dace may also have interbred with other species of dace, and collecting for scientific purposes probably also helped to reduce numbers. However, it seems that the most important factor in its extinction was the decision to allow a hotel to chlorinate hot spring water and discharge it into the marsh. The chlorine reacted to form chlorinated hydrocarbons, which are very toxic to fish, even at low concentrations. Shortly thereafter, the Banff longnose dace disappeared forever.

Migratory Bird Sanctuaries and National Wildlife Areas

Following the Migratory Birds Convention Act in 1917 Migratory Bird Sanctuaries were set aside to protect migratory bird species against hunting. Fifty years later, however, habitat loss had become a more significant threat to many populations and as a result another category of protected area, National Wildlife Areas (NWAs), was established under the Canada Wildlife Act and administered by the Canadian Wildlife Service. Section 8 of the Act specifically allows land to be protected for endangered species, as well as for migratory species. There are now over 150 such sites scattered across Canada; together with Migratory Bird Sanctuaries they cover over 11.8 million hectares. The Migratory Bird Sanctuary regulations prohibit all disturbance, hunting, and collection of migratory birds and their eggs. The regulations only control activities within sanctuary boundaries when migratory birds are actually present and do not provide protection of habitat. In contrast, National Wildlife Area regulations provide for both habitat protection and regulation of activities throughout the year.

These sites are of critical importance for some wildlife species, including species at risk. The piping plover, an endangered species, has nesting habitat protected in Pointe de l'Est NWA in Quebec. The new Suffield NWA in Alberta, with its short-grass prairie and sand dunes, is home to the burrowing owl (endangered) and ferruginous hawk (vulnerable). The proposed Isabella Bay NWA is home to part of the endangered Arctic population of bowhead whales.

Provincial Designations

Each province differs in terms of its system of protected areas. All have some form of provincial park system, although they differ significantly in terms of classifications and policies. Some allow extractive harvesting such as logging, trapping, and hunting. Furthermore, rarely in the past have such systems been established with ecological criteria central in determinng location, size, and boundaries, so the protection afforded endangered species can be quite uneven. Nonetheless, provincial parks can be large and afford valuable protection to some endangered species. In northwest Ontario, for example, Woodland Caribou Provincial Park is one of the few wilderness-class provincial parks, where extractive activities are prohibited and the forces of nature are allowed to run their course. It, along with adjoining Atikaki Wilderness Park in Manitoba, provides valuable habitat for the endangered woodland caribou.

Most provinces also have other forms of protective legislation that might include ecological reserves, wildlife management areas, and similar designations. Typically, such areas are small but have strong protective powers, and their location and boundaries have often been designed with ecological criteria as a main concern and endangered species an important factor. The Vancouver Island marmot, for example, is Canada's only endemic endangered mammal species, and one of the remaining colonies is protected by a provincially designated ecological reserve.

Under provincial wildlife statutes or in provincial endangered species legislation in the provinces that have it (see Chapter 3), there is often provision for areas considered to be essential for the preservation of endangered species to be set aside. The Vancouver Island marmot, for example, is protected in other colonies through their designation as Critical Wildlife Areas under the Wildlife Act. This is the only such designation in the province, however. The weaknesses of this patchwork provincial approach are further illustrated by Quebec, where recent changes to the Wildlife Act make protection of endangered species habitat more discretionary.

Private Stewardship

Although Canada has made significant progress in extending the size of its protected areas over the last decade, a considerable proportion of the habitat needed to safeguard endangered species is in private hands. Consequently, increasing effort is being made to encourage private owners to protect such habitat and to compensate them for losses that might be incurred due to such measures. For example, in Ontario the Conservation Land Tax Incentive Program is designed to recognize, encourage, and support the long-term stewardship of specific classes of conservation lands by offering tax relief to landowners who agree to participate in the program. Habitats of species protected under the Endangered Species Act are eligible under the program and include listed species such as bald eagle, peregrine falcon, piping plover, loggerhead shrike, henslow's sparrow, blue racer and Lake Erie water snakes, and the cucumber tree.

Not all programs provide such financial incentives. In Alberta, for example, the Operation Burrowing Owl (OBO) program of the Alberta Fish and Game Association enlists farmers, ranchers, and other rural landowners in southern Alberta as volunteers in the protection of active and previously active burrowing owl nesting areas. Participating landowners sign a voluntary agreement to preserve owl nesting sites for a period of five years. Through such arrangements, 243 members of OBO are currently protecting 26,371 hectares of grassland habitat in Alberta. Suggestions for habitat preservation or improvement are provided in annual newsletters mailed to all members.

THREATS FACING ENDANGERED SPECIES PROTECTION IN TERRESTRIAL PROTECTED AREAS

There are many different ways to categorize threats to endangered species in protected areas. One of the most common is whether the threat originates inside or outside the area boundary. At first, little attention was given by protected area management to events outside their administrative boundaries. Now, however, protected areas are recognized as but one piece of the landscape matrix and equally vulnerable to many of the same factors responsible for habitat degradation on the rest of the landscape.

Internal Threats

Visitation

Although national parks and most provincial parks, at least in theory, give priority to ecological integrity in management decisions, they also have a mandate to provide for the enjoyment of visitors. Indeed, since the inception of parks in Canada, this aspect of the mandate has garnered most of the management attention (Dearden and Berg, 1993) and visitor numbers have soared, often to the detriment of various species.

In 1978 the piping plover was designated by COSEWIC as threatened in Canada. The piping plover is a small shorebird that nests in shallow scrapes on undisturbed beaches. The nests are very vulnerable to weather variables, predators, and human disturbance. Since 1991, plover populations have declined by about 17 per cent in Atlantic Canada. Breeding populations occur in three national parks, Prince Edward Island, Kouchibouguac, and the Seaside Adjunct to Kejimkujik. Early reports on the status of the plover indicated that human disturbance could be a critical factor (Martin and Cairns, 1979). Management prescriptions meant to reduce disturbance were not strictly enforced and numbers continued to decline until 1985, when COSEWIC upgraded the plover to endangered status.

Since that time a more serious management approach has been taken towards protection of the species, and since 1988 all three parks have provided close monitoring and enforcement of regulations, closure of nesting areas to the public, and protection of individual nests from predation using exclosures. In general, these measures have been successful. The recovery plan goal of over 1.5 chicks per breeding pair has been met overall within the parks. However, within the Prince Edward Island park nesting and hatching success was found to be higher in the period before strict management measures were undertaken. This decrease is credited largely to increases in nest depredation, despite the exclosures, and Corbett (1997) suggests that an increased predator population may be to blame, with increases attributed to development and habitat modification outside the park.

One important lesson from the piping plover experience emphasizes that species at risk are often vulnerable to several different pressures. These can be both direct, as in the case of direct human disturbance of breeding behaviour, and indirect, where human alterations lead to changes in relative species abundances that have unfavourable consequences for some species through food chain interactions. In this case, increased human fragmentation of habitat led to increases in human-tolerant predators such as raccoons, skunks, gulls and crows.

Development

It is not just the numbers of visitors that place pressure on wildlife, but also the services constructed for those visitors. The 1997 *State of the Parks Report* (Minister of Public Works and Government Services Canada, 1998) reported visitor and tourism facilities as being the most frequent stress to park ecosystems. Perhaps no area has sustained more pressure in this regard than the Banff-Bow Valley region. The town-site itself is situated in one of the most productive and rare habitats throughout the Rocky Mountains, the Montane zone, which covers less than 5 per cent of the park area. This zone has frequent chinook winds, low snow accumulation, warm winter temperatures, migration routes, and diverse habitats. These conditions are favourable for a dense concentration of wildlife and the zone has critical wildlife areas. It is also very attractive for development purposes. Over 70 per cent of the zone has already been occupied by highways, golf courses, towns, and resorts—and more development is planned.

One species not tolerant of this level of disturbance is not only critical to the ecosystem but also an endangered species, the grizzly bear. Grizzly bear mortalities in Banff are in fact higher than outside the park. There has been a steady erosion of griz-zly bear habitat within the park and increased fragmentation of core habitat due to developments (Gibeau et al., 1996). As a result the bears have come into increasing conflicts with humans. Mortality analysis revealed that problem wildlife control actions accounted for 71 per cent of grizzly bear mortalities, followed by highway and railway kills (17 per cent), unknown (8 per cent), and natural death (3 per cent). Over 90 per cent of grizzly bear mortalities in Banff Park occurred in front-country areas, within a 500-metre zone surrounding roads and human infrastructure (Benn, 1998). As a result of the careful analysis of this situation by a multi-stakeholder research partnership, described by Herrero et al. (1998), policy-makers curtailed future devel-opments in the area that would have a further negative impact on bear populations.

Transport
Transportation routes such as roads and railway lines are one of the greatest contrib-utors to habitat fragmentation, which is one of the greatest causes of species declines. Although road density is less within than outside parks, wildlife in older parks in particular might suffer from the effects of roads, both directly in terms of mortality and also indirectly in terms of the other disturbances that are facilitated once road access is gained.

Perhaps the best-known example of the impacts of roads is the Trans-Canada Highway as it cuts through Banff National Park. The road counts for about half the reported wildlife deaths in the park (Shury, 1996), and mortality levels within the park for some wildlife populations are equal to or greater than hunted populations outside the park (Gibeau and Heuer, 1996). Both underpasses and overpasses have been con-structed and the roads have been fenced to try to mitigate the impacts. These meas-ures seem to have worked well with ungulate populations, i.e., hoofed animals (Woods, 1990), but their efficacy with the more endangered grizzly and wolf popula-tions is still under investigation.

In light of the well-known detrimental impacts of roads on biodiversity and the policy of the national parks against constructing any new roads, it is difficult to understand the preliminary approval in late 1999 from the federal Heritage Minister,

Sheila Copps, to construct a new all-weather road through Wood Buffalo National Park. The formal rationale for the road to Fort Smith is to increase tourism traffic. However, the community is already accessible by an all-weather road, and the new road will destroy one of the main things that the park was established to protect— wilderness. It will also facilitate movement of COSEWIC-designated threatened bison out of the park and add to concerns over the spread of diseases such as brucellosis to cattle in Alberta.

Harvesting
Under the National Parks Act commercial extraction of resources from parks is not allowed. However, in many national parks, especially in the North, traditional harvesting activities have been allowed to continue, subject to certain conditions. Most of these agreements are for First Nations people, but agreements also exist that allow, for example, duck hunting in the Mingan Archipelago Reserve National Park along the north shore of the Gulf of St Lawrence in eastern Quebec and snowshoe hare snaring in Newfoundland's Gros Morne National Park.

The impact on the parks and park wildlife of the Supreme Court ruling in September 1999 that gave Mi'kmaq fishermen in Atlantic Canada the right to earn a 'moderate livelihood' through year-round hunting, fishing, gathering, and trading is difficult to judge. However, some interpretations have suggested that it may result in year-round hunting throughout Canada's protected areas, including the national parks. Already in Alberta there are challenges to the province's hunting laws by the Native community.

Good justification can be found for most of the extractive activities still allowed. In many cases, failure to allow such activities, especially in northern Canada, would have negated the possibility of park establishment. Furthermore, these activities are confined to existing extractive activities and would not be allowed on any species at risk. Nonetheless, it should be borne in mind that parks have a critical role in terms of their 'benchmark' function. The further we depart from the pristine, the less likely it is that they can fulfil this role adequately.

Illegal harvesting or poaching is also a concern. Statistics are difficult to obtain, but the Canadian Wildlife Federation estimates that double the numbers of animals are killed illegally as are taken legally in Canada. In the four mountain parks (Banff, Jasper, Yoho, Kootenay) at least 60 animals have been illegally killed over the last decade. Fines for violations were raised to a maximum of $150,000 in the 1988 amendments to the National Park Act. However, the chances of being caught are still minimal and the rewards for poaching are increasing, especially in response to demands from the Asian market. Sometimes the poaching victims are of special concern, such as the American marten juvenile, found in an illegal snare, which was one of the offspring of two that had been reintroduced into Terra Nova National Park in Newfoundland.

One aspect of harvesting in parks that until recently was accorded little ecological attention is recreational fishing. Fishing has been actively encouraged in most parks, and only in Forillon National Park, on the Gaspé Peninsula, is it banned. Given that aquatic ecosystems contain a high proportion of endangered species in Canada and that our knowledge of such systems is limited, a more prudent approach to recreational fishing is called for in the future.

External Threats

External threats are now recognized as the most serious causes of loss of ecological integrity in most protected areas. The proximity to the protected area, the nature of the threat, and its intensity vary widely. At the local scale, for example, land-use practices just outside the boundary can have a negative impact on species inside the protected area. The ability of park management authorities to influence such impacts varies. In 1996, for example, concern over declining wolf populations led to a request by Riding Mountain National Park to Manitoba's Department of Natural Resources to remove wolves from the list of species that could be hunted around the park. As a result, wolf hunting was dropped from 10 to three months of the year, with also a reduction in wolf trapping.

Other impacts are much less amenable to change at the local level. The pollutants that threaten the health of the endangered beluga whales at Saguenay National Park in Quebec originate in industrial areas many kilometres away. Other pollutants are airborne. Samples taken of the snow pack in the Rocky Mountain parks—sites that we think of as being among the most pristine on earth—show higher concentrations of many pesticides than near-urban and agricultural sites (Blais et al., 1998).

As noted earlier, a high proportion of species at risk are found in Canada's national park system. The same can also be said for the numbers of exotic species (i.e., species introduced deliberately or inadvertently by humans to an ecosystem outside its normal range). About half of Canada's exotic plant species are present in parks, as are 79 per cent of exotic vertebrates (Minister of Public Works and Government Services Canada, 1998). Gwaii Hanaas in British Columbia is one park whose ecosystems have suffered quite dramatically as a result of exotic species such as Sitka black-tailed deer, raccoons, squirrels, beaver, and three species of rats. The rats and raccoons have been particularly detrimental to the burrow-nesting seabirds for which the archipelago is so well known. On Kunghit Island, for example, the ancient murrelet colony decreased in size by approximately one-third between 1986 and 1993 as a result of depredation by introduced Norway rats. Gaston (1994) suggests that unless the spread of raccoons can be halted and rats eliminated on these islands the murrelet population will be extirpated in the near future.

Not all problems with exotics are confined to animals. The red mulberry is a rare tree found in the Carolinian region of southern Ontario, where its numbers are declining due to habitat loss and hybridization with the non-native white mulberry. As a result, action is under way at Fish Point Provincial Nature Reserve and at Point Pelee National Park to locate, cut, and spray the white mulberry to prevent further hybridization as part of the first recovery plan for a tree species in Canada.

FUTURE CHALLENGES

Need for More Protected Areas

Protected areas play a fundamental role in trying to maintain and enhance populations of many endangered species. Habitat degradation is the greatest causal factor
behind endangerment at virtually every scale. Setting aside areas to try to minimize
such degradation is a basic necessity if we are to slow down the slide of many species
into the endangered categories and eventually into extinction. This will require setting
aside more of the landscape into various categories of protected area. We currently
have 39 terrestrial national parks, but only 24 of Parks Canada's natural regions are
represented, despite a political commitment to complete the system by the year 2000
(Environment Canada, 1990).

Some provinces have made great strides in terms of expanding the amount of
area protected over the last decade. BC has the largest proportion of the land base
protected and other provinces, such as Ontario, Manitoba, and Nova Scotia, have
added large amounts of land. Others, however, have remained virtually static.
Alberta has added about 1 per cent while Quebec, New Brunswick, and
Newfoundland have contributed even less. The Ontario Forest Accord saw the addition of 378 new and expanded protected areas, totalling 2.4 million hectares, while
Nova Scotia designated 31 new protected areas covering 300,000 hectares. Despite
these increases it is apparent that Canada still has a long way to go towards establishing a system of protected areas that will provide adequate habitat protection for
many species.

Big Is Good, Linked Is Better

Different species have different habitat requirements. It is difficult to plan for such
diversity in designing protected area systems. Certain species, known as 'umbrella
species', may have requirements that, if met, will also shelter many other species. Large
carnivores, such as grizzly bears, are one example, as they have large area requirements
and dispersal distances. To protect such species, protected areas have to be large—very
large, in the order of thousands of square kilometres (Noss et al., 1996). Although
Canada has some vast reserves, most of the large ones are in the North, whereas most
of the pressure on endangered species is in the south, where reserves are small
(Minister of Public Works and Government Services Canada, 1998).

In terms of maintaining viable populations of many species, attention has thus
changed not just to having large reserves, but also to linked networks of reserves
(Noss and Harris, 1986; Paquet and Hackman, 1995). Some of the better-known
examples are the Yellowstone to Yukon (Y to Y) campaign of the Canadian Parks and
Wilderness Society and the more recent eastern initiative, the Adirondack to
Algonquin (A to A). Both attempt to link existing core reserves with new reserve
proposals and corridors through multiple-use landscapes that will facilitate the
physical and genetic interchange of various species through the larger landscape
matrix.

Need for Greater Attention to the Ecological Role of Protected Areas

Besides having more protected areas and linking them through networks, it is also essential that full attention at the political and management levels be devoted to ensuring that they fulfil their ecological mandates. Many politicians and even managers are still challenged by the thought that in protected areas the requirements of other species should always be given priority over the business and recreational interests of humans. Although Parks Canada ostensibly is the most advanced jurisdiction in making this adjustment, the release of the report by the 'Panel on the Ecological Integrity of Canada's National Parks' (Parks Canada Agency, 2000) clearly demonstrated that even in the national park system there are significant problems to be overcome before the ecological integrity mandate can be achieved. One fundamental necessity is to have rigorous, systematic, and comprehensive monitoring systems established (Woodley, 1994).

Lack of Understanding about the Species

Our knowledge of the ecology of many species, including endangered species, is still sadly lacking. If we do not know what species require, it is obviously difficult to formulate recovery plans for them (Foin et al., 1998). One fundamental role of national parks is to provide 'outdoor laboratories', where scientists can learn more about natural ecosystems. For example, the Blanding's turtle was designated as threatened by COSEWIC in 1995, with limited habitat, low populations, unnaturally high predation rates, limited recruitment, and human disturbance being the main areas of concern. As a result, scientists from Acadia University, the World Wildlife Fund, and the Nova Scotia Department of Environment, along with park scientists, have been undertaking studies on turtle ecology in Kejimkujik National Park. The scientific information gathered to date has filled in many gaps and forms the core of the species management plan for use both inside and outside the park (Herman et al., 1998).

Lack of information is ubiquitous, but it is an especially serious concern for freshwater species, where in many instances few systematic ecological data are available. Ricciardi and Rasmussen (1999) point out that since 1900, 123 freshwater animal species have been recorded as extinct in North America. There are currently 92 species of fish, molluscs, and amphibians on the COSEWIC lists and Ricciardi and Rasmussen suggest that extinction of freshwater animals is at rates equal to those for species native to tropical forests. Massive habitat deterioration is cited as the main cause, and given the responsibility of protected areas to halt such deterioration, better information on species and habitat requirements is clearly required.

Need for an Ecosystem-based Management Approach

From the above discussion of external threats it should be apparent that most protected areas face severe challenges to their ecological integrity from forces outside their administrative boundaries. Furthermore, as ecological knowledge improves we are becoming increasingly aware of the links between different species and between species and their environments, and of the influence of humans on the environment. Such awareness has focused greater attention on ecosystem processes rather than simply on components. In the context of endangered species, this means realizing that species cannot be helped to recovery without due attention being paid to the ecosystem conditions within which they live, as opposed to merely focusing on the numbers of the species. This concern for addressing the broader context has been termed ecosystem management (Grumbine, 1997; Slocombe, 1998) and is the management approach officially adopted by Parks Canada (Ministry of Supply and Services, 1994). Nelson and Skibicki (1998) provide an example of how ecosystem-based planning was used in Pukaskwa and Georgian Bay Islands National Parks, Ontario.

Need for Greater Federal-Provincial Co-operation
One of the most fundamental weaknesses in the Canadian approach to resource management is the general antipathy between the different levels of government and the challenges in getting them to work together co-operatively for the good of all Canadians and the Canadian landscape. This is also true with federally designated protected areas and the surrounding lands under provincial jurisdiction. Traditionally, federal authorities have been loath to advocate any management strategies on lands outside their jurisdiction, even if proposed land uses would result in a detrimental impact with the protected area. This has often been true, for example, with logging in provincial forests on the boundaries of national parks (Dearden and Doyle, 1998).

This has to change if an ecosystem-based approach is going to be implemented. And there are signs that it is changing. In Alberta, where the antipathy between the two senior levels of government can be quite intense, Banff park officials have made it known that they will be testifying on the likely impacts of the proposed Spray Lakes development in adjacent Kananaskis Country. The proposed development, by Calgary-based Genesis Land Development Corp., would include a year-round resort, 27-hole golf course, room for 6,000 skiers a day, 400 accommodation units, a helicopter ski operation, and a tour boat business. However, the area is considered critical habitat for grizzly bears. Canadian national parks by themselves only protect approximately 3.4–4.4 per cent of Canada's grizzly bears so inter-agency co-operation is essential to protect viable populations in larger grizzly bear ecosystems (Herrero, 1994).

Need for Active Management for Many Species and Parks
Protected areas are part of the dynamic landscape pattern, as emphasized above, and although we can try to protect them from undesirable outside influences, it is impossible to insulate them completely from these changes. Thus, the exotic species described earlier will readily invade the parks if allowed to do so. Furthermore, we are also learning that many ecosystems adapt to disturbances over time. Fire, for example, was a main ecological feature of many park ecosystems before we tried to halt such dynamism through fire suppression policies. Both the need to counter undesir-

able ecological changes invading protected areas from the rest of the landscape and the realization of the necessary dynamism of ecosystems have led to much wider acceptance of the need for active management of protected areas. Parks Canada, for example, has a fire restoration program across Canada. Other examples of active management include the headstart program for the Blanding's turtle in Kejimkujik and the efforts to eliminate alien predators such as rats and raccoons from some islands in Gwaii Hanaas National Park Reserve.

Active management is contrary to the traditional 'hands-off' approach to park management. However, times have changed. In analysing recovery plans for endangered species in the US, Foin et al. (1998) found that in 63 per cent of the over 300 recovery plans available, active management was a necessity, with mere habitat protection, the forte of protected areas, no longer sufficient. The Ecological Society of America (Carroll et al., 1996) reached similar conclusions. For many managers this need to anticipate and react in a proactive way in areas we have traditionally left alone is going to be a challenge.

Finally, it should be noted that although much of this chapter has concentrated on wildlife and the role of large, natural protected areas, in fact, there are more plants than mammals, birds, or reptiles on the national list of species at risk. Furthermore, although large natural protected areas are also important to plant protection, so, too, are many botanical gardens and arboreta that are involved in endangered plant recovery programs. These gardens are run by various levels of government, universities, and private institutions, and their role mirrors future challenges for protecting endangered species by terrestrial protected areas. We will see an increasing range of actors involved in a variety of co-operative activities to help forestall declines and rebuild populations of many species in both *in situ* and *ex situ* locations at all scales.

REFERENCES

Auditor General. 1996. 'Canadian Heritage—Parks Canada: Preserving Canada's Natural Heritage', ch. 31 in *Report of the Auditor General of Canada*. Ottawa.

Benn, B. 1998. 'Grizzly Bear Mortality in the Central Rockies Ecosystem, Canada', Master's Project, University of Calgary.

Blais, J.M., D.W. Schindler, D.C.G. Muir, L.E. Kimpe, D.B. Donald, and B. Rosenberg. 1998. 'Accumulation of persistent organochlorine compounds in mountains of western Canada', *Nature* 395: 585–8.

Carroll, C., C. Auspurger, A. Dobson, J. Franklin, G. Orains, W. Reid, R. Tracy, D. Wilcove, J. Wilson, and J. Lubchenco. 1996. 'Strengthening the use of science in achieving goals of the Endangered Species Act', *Ecological Applications* 6: 665–91.

Corbett, G.N. 1997. *Conservation of the Endangered Piping Plover* (Charadrius melodus) *in Canada's National Parks*. Halifax: Parks Canada.

Dearden, P., and L. Berg. 1993. 'Canada's National Park: A Model of Administrative Penetration', *Canadian Geographer* 37: 194–211.

—— and S. Doyle. 1998. 'External Threats to Pacific Rim National Park Reserve, BC', in C. Stadel, ed., *Themes and Issues of Canadian Geography II*. Salzburger Geographische Arbeiten, 121–36.

Environment Canada. 1990. *Canada's Green Plan*. Ottawa: Supply and Services.

Foin, T.C., S.P.D. Riley, A.L. Pawley, D.R. Ayres, T.M. Calsen, P.J. Hodum, and P.V. Switzer. 1998. 'Improving recovery planning for threatened and endangered species', *Bioscience* 48: 177–84.

Gaston, A.J. 1994. 'Status of the ancient murrelet (*Synthliboramphus antiquus*) in Canada and the effects of introduced predators', *Canadian Field-Naturalist* 108: 211–22.

Gibeau, M.L., S. Herrero, J.L. Kansas, and B. Benn. 1996. *Grizzly Bear Population and Habitat Status in Banff National Park: A Report to the Banff-Bow Valley Task Force*. Calgary: Eastern Slopes Grizzly Bear Project, University of Calgary.

Grumbine, R.E. 1997. 'Reflections on "What is ecosystem management"', *Conservation Biology* 11: 41–7.

Herman, T., J. McNeil, N. McMaster, and I. Morrison. 1998. 'Recovery of a threatened Blanding's turtle population: Linking conservation efforts in working and protected landscapes', in N.W.P. Munro and J.H.M. Willison, eds, *Linking Protected Areas with Working Landscapes Conserving Biodiversity*. Wolfville, NS: Science and Management of Protected Areas Association.

Herrero, Stephen. 1994. 'The Canadian National Parks and Grizzly Bear Ecosystems: The Need for Interagency Management', *International Conference on Bear Research and Management* 9, 1: 7–21.

——, Jillian Roulet, and Michael L. Gibeau. 1998. 'Banff National Park: Science and Policy in Grizzly Bear Management', *International Conference on Bear Research and Management* 11 (forthcoming).

Martin, K., and W.E. Cairns. 1979. *Avifaunal Survey of Prince Edward Island National Park*. Ottawa: Canadian Wildlife Service.

McNamee, K. 1993. 'From Wild Places to Endangered Spaces: A History of Canada's National Parks', in P. Dearden and R. Rollins, eds, *Parks and Protected Areas in Canada: Planning and Management*. Toronto: Oxford University Press, 17–44.

Minister of Public Works and Government Services Canada. 1998. *State of the Parks: 1997 Report*. Ottawa.

Nelson, J.G., and A.J. Skibicki. 1998. 'Some Approaches to Planning for Greater Park Ecosystems in Ontario, Canada', in C. Stadel, ed., *Themes and Issues of Canadian Geography II*. Salzburger Geographische Arbeiten, 9–23.

Noss, R.F., and L.D. Harris. 1986. 'Nodes, networks and MUMS: Preserving biodiversity at all scales', *Environmental Management* 10: 299–309.

——, H.B. Quigley, M.G. Hornocker, T. Merrill, and P.C. Paquet. 1996. 'Conservation Biology and Carnivore Conservation in the Rocky Mountains', *Conservation Biology* 10: 949–63

Panel on the Ecological Integrity of Canada's National Parks. 1999. *Unimpaired for Future Generations? Conserving Ecological Integrity with Canada's National Parks*. Ottawa: Report to Minister Copps.

Paquet, P.C., and A. Hackman. 1995. *Large Carnivore Conservation in the Rocky Mountains*. Toronto: World Wildlife Fund.

Parks Canada. 1986. 'The designation and management of rare, threatened and endangered species', Management Directive 2.4.3. Ottawa: Natural Resources Division.

——. 1994. *Guiding Principles and Operational Policies*. Ottawa: Ministry of Supply and Services.

Parks Canada Agency. 2000. '*Unimpaired for Future Generations'? Protecting Ecological Integrity with Canada's National Parks*. Report of the Panel on the Ecological Integrity of Canada's National Parks. Ottawa.

Ricciardi, A., and J.B. Rasmussen. 1999. 'Extinction Rates of North American Freshwater Fauna', *Conservation Biology* 13: 1220–2.

Shury, T. 1996. *A Summary of Wildlife Mortality in Banff National Park, 1981–1995*. Final report submitted to the Warden Service, Banff National Park, Alberta.

Slocombe, D.S. 1998. 'Defining goals and criteria for ecosystem-based management', *Environmental Management* 22: 483–93.

Wipond, K., and P. Dearden. 1998. 'Obstacles to Maintaining Ecological Integrity in Pacific Rim National Park Reserve', in N.W.P. Munro and J.H.M. Willison, eds, *Linking Protected Areas with Working Landscapes Conserving Biodiversity*. Wolfville, NS: Science and Management of Protected Areas Association, 901–10.

Woodley, S. 1994. 'A scheme for ecological monitoring in national parks and protected areas', *Environments* 23: 50–73.

Woods, J. 1990. *Effectiveness of Fences and Underpasses on the Trans-Canada Highway and Their Impact on Ungulate Populations Project*. Final Report. Prepared for Canadian Parks Service, Banff National Park Warden Service.

Endangered Marine Species and Marine Protected Areas in Canada

Martin Willison

Marine Species at Risk

Until recently, it was widely considered that human-caused extinction was not likely to occur in the marine realm. There has been a persistent, but false, belief that marine populations and ecosystems are 'resilient' and that human impacts in the oceans are negligible. For many people, however, recent fishery collapses have cast doubt on this widespread confidence in the capacity of marine systems to rebound. Human impacts on the continental shelves in particular are now known to be substantially greater than was generally realized, with some 25–35 per cent of primary productivity (products of photosynthesis) already being consumed by human activity (Pauly and Christensen, 1995). Fisheries have substantial impacts on non-target species in the form of bycatch and incidental habitat damage (for review, see Dayton et al., 1995). As a result of over-exploitation in fisheries, humans are eating progressively 'down the marine food web' (Pauly et al., 1998). The world's continental shelves are swept on average by trawls at least once every two years (Safina, 1998), and in many places the trawling intensity is particularly great, a practice that has been likened to forest clear-cutting (Watling and Norse, 1998). In practice, anthropogenic extinction of species and populations has been a real, but relatively unnoticed, phenomenon in the world's seas and oceans (for review, see Roberts and Hawkins, 1999; Carlton et al., 1999).

Of the 328 extant populations and species listed as 'wildlife at risk' by World Wildlife Fund Canada in 2000, only about 8 per cent are marine. The majority (21) of these are marine mammals. Of these 21 mammals, 17 are cetaceans (whales and their close relatives), constituting almost 5 per cent of the total wildlife at risk. Globally, several cetacean populations have been extirpated in the last 200 years, and a few species are critically endangered, but no cetacean species has become extinct (Carlton et al., 1999). In contrast, only one strictly benthic (bottom-dwelling) marine species is listed by WWF Canada: the northern abalone (a mollusc). Despite this, there were several probable extinctions of marine invertebrate animals in the twentieth century (Roberts and Hawkins, 1999). This suggests a bias towards the listing of charismatic marine megafauna that appear at ocean surfaces.

In creating its list, World Wildlife Fund uses the designations of the Committee on the Status of Endangered Wildlife in Canada (COSEWIC), the World Conservation Union's Red Data list, and those species listed in the United States under the Endangered Species Act, where the same species is found in Canada and is judged to be at risk in Canada. The processes used by the listing committees are outwardly rigorously scientific, but this does not overcome the tendency to pay special attention to species having a popular constituency.

Birds and mammals are the best-known groups of organisms biologically, with all species in the world named and described. Thus, it is not surprising that they constitute the bulk of wildlife considered at risk, for knowledge is required to estimate the status and vulnerability of a species. In the marine realm, however, most non-microscopic species live on the bottom and are unfamiliar, with many species unknown to science. For example, sampling in 1997 of seamounts at depths of 660–2,000 metres to the south of Tasmania (Australia) revealed that about a third of all the invertebrate species found were of previously undescribed types (Koslow and Gowlett-Holmes, 1998). Many of these came from isolated ecosystems dominated by deep-sea corals that had been severely ravaged by the scouring effect of bottom trawling, suggesting that many species that existed on these seamounts may have been lost completely before there was an opportunity to describe them. Similarly, Carlton et al. (1999) estimate that tropical coral reef damage may have led to at least a thousand recent unnoticed extinctions, most of these being invertebrate animals.

It is a popularly held misconception that most marine animals range widely and are therefore less likely to be extinguished by intense human activity in a particular region. In their review of the risk of extinction in marine environments, Roberts and Hawkins (1999) concluded that 'small geographic ranges are common, especially among small-bodied organisms.' As an example, Wilson and Kaufman (1987, cited in Koslow and Gowlett-Holmes, 1997) estimated that about 15 per cent of the invertebrate species of any seamount or seamount group are endemic to it. In other words, they are found there and nowhere else.

For the reasons outlined above, I will consider here examples of species that are formally acknowledged as being at risk in Canada, as well as examples of species and natural assemblages perched precariously within this confusion of ignorance and presumed immunity. Although I examine a large number of cases below, this is not a comprehensive review—several listed marine species are not mentioned here, and a very large number of unlisted species might also be considered. Similarly, I have selected among the marine protected areas nominated and established in Canada. I have tried to select examples that provide lessons of significance and, collectively, provide a coherent story. My selections are biased towards the Atlantic region of Canada, with which I am most familiar.

MARINE PROTECTED AREAS

Setting aside protected areas to provide habitat for all populations representative of a region is the primary strategy in any well-made conservation plan (Soulé, 1991). Such a strategy provides insurance against mistakes in natural resource management and underpins all secondary and tertiary conservation strategies. Systems of terrestrial protected areas have been developed in almost all jurisdictions in the world, but protected areas have not yet been established in the marine environment with equivalent zeal.

The World Conservation Union (IUCN) formally defines a marine protected area (MPA) as:

> Any area of intertidal or subtidal terrain, together with its overlying water and associated flora, fauna, historical and cultural features, which has been reserved by law or other effective means to protect part or all of the enclosed environment. (Kelleher and Kenchington, 1992)

Just as terrestrial protected areas fall into several categories, ranging from strict protection to relatively minor limitations on use, so MPAs also vary according to the degree of protection afforded. Strictly protected MPAs are relatively rare. For example, marine sanctuaries in the United States only restrict fishing in very small parts of the sanctuaries (Barr, 1995), and in Australia only about 7 per cent of MPA areas has been set aside for the protection of biodiversity (Ottesen and Kenchington, 1995).

As recently as 1991, Prince Charles, as president of the UK Marine Conservation Society, stated the following in the Foreword to *Marine Reserves for New Zealand* (Ballantine, 1991):

> There can be few countries in the world which do not possess, and protect, nature reserves on land, but there are only a handful of marine reserves in existence worldwide. . . . I believe the time has come to start looking very carefully at the need for such reserves. The principle is entirely straightforward; that there should be areas set aside in which no human disturbance is allowed. Provided that the areas are carefully chosen, and sufficiently large, natural levels of marine life can be protected and sustained.

MPAs of this sort are now often called 'no-take' reserves. Bohnsack (1993) has stated that such reserves 'enhance fisheries, reduce conflicts and protect resources'. Thus, in addition to marine reserves having been set aside for the purpose of protecting biodiversity, MPAs have been created for fishery management purposes (e.g., as sources of larval export to enhance recruitment to fisheries), to promote tourism (e.g., marine parks), to reduce conflicts among classes of users (e.g., multiple-use marine management areas), and to protect endangered species (e.g., whale sanctuaries). As a result of these many purposes and the diversity of legislative mechanisms created in various parts of the world, the terminology used to classify MPAs is very confusing. For this reason, the term 'marine protected area' has been adopted as an umbrella (for detailed reviews on marine protected areas, see Gubbay, 1995; Kelleher, Bleakley, and Wells, 1995; Agardy, 1997).

MARINE PROTECTED AREAS IN CANADA

The marine environment falls under federal jurisdiction in Canada. MPAs have been established by several means, and new legislative and regulatory mechanisms are still being developed. The Fisheries Act contains provisions for habitat protection that have been used to establish de facto MPAs by excluding fishing from specified areas for long periods (e.g., the Western and Emerald Banks closure on the Scotian Shelf). The Parks Act has been used to establish national 'marine parks'. These are managed by the Department of Canadian Heritage (Parks Canada) and emphasize interpretation for visitors. The Parks Act was designed for terrestrial parks, however, and is poorly suited for use in protecting marine areas.

The Oceans Act, which became law in 1997, has strong provisions for the establishment of MPAs, and requires the Minister of Fisheries and Oceans to create them ('shall' is used in the legislation, not 'may'). Several Oceans Act 'pilot MPAs' have been announced, although regulations and management plans for these have not yet been created. Oceans Act MPAs may be established for several purposes, but not for the protection of marine biodiversity through the systematic representation of marine natural regions in a network of protected marine wilderness areas.

At the time of writing, a bill to permit the establishment of 'National Marine Conservation Areas' (NMCAs) for the purpose of protecting representative elements of marine biodiversity is being considered. The Department of Canadian Heritage (Parks Canada) would be the responsible agency for NMCAs, and it is expected that Parks Canada will work closely with the Department of Fisheries and Oceans (DFO) in managing these. The legislation has had a very long gestation (Parks Canada, 1994, 1995).

Remarkably, only four of the 75 established Canadian MPAs listed in the authoritative review *A Global Representative System of Marine Protected Areas* (Kelleher, Bleakley, and Wells, 1995: vols 1 and 4) are of any of these types. The remainder include 33 migratory bird sanctuaries established under the Migratory Birds Convention Act, seven National Wildlife Areas established under the Canada Wildlife Act, and a variety of provincial parks and reserves that have no clear legal mandate below the high-water mark. The Migratory Bird Sanctuaries and National Wildlife Areas are classified as IUCN Category IV protected areas and are managed by Environment Canada. Hunting of migratory birds and disturbance of nests are prohibited in the bird sanctuaries, but other exploitive activities may continue, provided a permit is obtained. The wildlife management areas protect coastal birds, particularly ducks and geese, and mammals, notably polar bears (vulnerable) and walruses. Permission may be granted for any activity within a wildlife management area, including hunting.

The level of protection afforded to obligatively aquatic species in Canada's marine protected areas is currently very low. Marine species that spend some time on land, or in the air, are more likely to be afforded protection than those that live entirely in water, particularly if the water is salty.

EXAMPLES OF LISTED EXTINCT AND ENDANGERED MARINE SPECIES IN CANADA

Two Auks

Examples of extinction and endangerment of marine species provide insights into the risks against which protection should be afforded. The renowned great auk (*Pinquinus impennis*), for example, was once widely distributed in the North Atlantic, but became extinct in the mid-1800s as a result of cumulative over-exploitation for several purposes. It was an agile bird in water but very vulnerable at the rocky islets on which it bred. Had a few of these islets been set aside, the species might be with us yet. A small auk of the Pacific, the marbled murrelet (listed by COSEWIC as threatened), is similarly vulnerable at its breeding sites (Rodway, 1990). It requires large old-growth moss-covered coastal trees for nesting, such as those found in the coastal temperate rain forests of Vancouver Island. Only a protected-area strategy can retain this critical habitat by reserving it against logging, the main threat. This alone may not be enough, however, for this auk also seems to be dying at an unsustainable rate in gill nets set to catch fish in the waters in which the marbled murrelet also fishes (ibid.).

Two Ducks

Another example of instructive extinction is provided by the Labrador duck, which was naturally rare and apparently had a low reproductive rate and specific habitat requirements for feeding. It probably became extinct in 1875 as a result of the cumulative effects of hunting, specimen collecting, and loss of prey (Kirk, 1985). The last known bird was shot. The eastern North American population of the harlequin duck has a population of about 1,000 and is endangered. The species has several of the same characteristics as the Labrador duck. During the breeding season it is a specialist feeder in rapidly flowing streams, and in winter it can be at risk of being shot when associating with other marine ducks (Montevecchi et al., 1995). In addition, it is threatened by oil spills, a problem not encountered by the Labrador duck. Species such as these have low tolerance for increased mortality. Harlequin duck breeding sites deserve strict protection, and where overwintering concentrations are found, notably in bays in the Gulf of Maine (Montevecchi et al., 1995), community-based conservation planning should be considered (as in the Important Bird Areas program of Birdlife International).

Two Ciscos and a Wolffish

Establishing permanent MPAs for harlequin ducks is an unlikely strategy because the ducks congregate unpredictably in their winter range. A fish once found in Lakes Huron and Michigan provides us with a possible example of the potential for application of an MPA strategy, however. The deepwater cisco (*Coregonus johannae*) was subjected to an aggressive commercial fishery until the 1930s, when its numbers became too low for a directed fishery. Although little is known about the biology of the species (Parker, 1988), its preference for deep water (50-160 metres), the existence of preferred fishing locations, and its former abundance all suggest that it might not have become extinct had some suitable pockets been set aside against fishing in the Great Lakes.

The status of the Atlantic cisco (*Coregonus huntsmani,* also called the Acadian white-fish), a relative of the deepwater cisco, tends to support the contention that fish species can benefit from a protected-area conservation strategy. This species is listed by COSEWIC as endangered, and healthy populations now exist in lakes associated with only one river in Nova Scotia, the Petite Riviere. The naturally anadromous species appears now to have been lost from other parts of its former range and the remaining population is landlocked by dams. Fortunately for the fish, the lakes occupied by the fish lie within the protected water-supply drainage for the town of Bridgewater (Edge, 1984). Provincial and municipal regulations protect the habitat and the species.

A vulnerable saltwater fish species, the Bering wolffish (*Anarhichas orientalis*), might also benefit by being provided with a marine protected area to secure its habi-tat (Houston and McAllister, 1989). Although there appears to be little pressure on this species at present in Canada, the Canadian population is found only in certain locations in Bathurst Inlet in Nunavut.

Comparison of the Atlantic cisco with other endangered species in the Atlantic region of Canada provides an interesting illustration of popular prejudice. This fish is the only Nova Scotian endangered species restricted entirely to the province, yet it is little known even in the region of its only refuge. There is no formal recovery plan for the species. The piping plover, a bird of sandy shores found during the summer in scattered locations on the Prairies, as well as in small numbers on beaches through-out Atlantic Canada and New England, is far better known and is an emblem for endangered species in the region. A great deal of effort is expended on its protection by federal, provincial, and voluntary agencies. Even Nova Scotia's vulnerable popula-tion of Blanding's turtle has a higher profile than the Atlantic cisco, but it is at risk only because its local population is small and isolated. Nova Scotia's Blanding's turtle and piping plover populations are deserving cases for concern, but on an even play-ing field of concern among biological taxa should they be more deserving than the local whitefish?

In his report to COSEWIC on the status of the endangered Atlantic cisco, Edge (1984) provides three reasons for regarding the species as specially significant: it is uniquely endemic to Nova Scotia; it is 'an exciting gamefish and tasty'; and it was fea-tured on a postage stamp. It is hard to imagine a critically endangered endemic bird or mammal being described in this way. Endangered birds and mammals are valued because they exist; no additional justification for their continuance is required. In the realm of species endangerment, it would seem that fish are second-class citizens, however, and the justification for their existence appears also to need an assessment of their palatability.

Eelgrass Limpet and Leatherback Turtle
Even for a species with small range requirements, we cannot expect that a protected-area conservation strategy can protect against all eventualities, as illustrated by the

eelgrass limpet. This small mollusc disappeared in 1929 throughout eastern North America as a result of the dramatic decline of eelgrass (*Zostera marina*) on which it fed. The eelgrass was devastated by the spread of a slime mould that consumed it, and although eelgrass has now recovered substantially, the recovery came too late for its associated limpet. It took 50 years for the extinction of the eelgrass limpet to be noticed (Carlton et al., 1999).

Similarly, protected areas cannot be set aside for species whose individuals treat the world's oceans as their backyard, such as the endangered leatherback turtle, which appears unpredictably in Canadian marine waters. It cannot be protected by setting aside marine waters for it in Canada, but instead we must focus on reducing the numbers caught incidentally on surface long-line hooks and other fishing gears.

Sea Otter

Fortunately, there are examples that provide hope for the recovery of endangered marine species. In the eighteenth century, the sea otter (*Enhydra lutris*) was widely distributed in the northern Pacific Ocean, but by 1911 its population had fallen to less than 2,000 individuals, surviving as 13 relict populations (Watson et al., 1996). It had been trapped to near extinction for an insatiable fur market. Following protection, some of the relict groups declined further, and by 1929 there were no sea otters in Canada (Kenyon, 1969). Between 1969 and 1972, 89 individuals were introduced to northern Vancouver Island, and after a short period of decline this translocated population began to grow by almost 20 per cent per year. The sea otter has since spread both northward and southward along the British Columbia coast, and the Canadian contingent now contributes to a healthy world count of some 150,000 individuals (Watson et al., 1996).

The government of British Columbia has contributed to the protection effort by declaring an ecological reserve in Checleset Bay to conserve sea otter habitat, as well as by regulating against trapping under its Wildlife Act. In practice, reserve status probably makes little difference, for the sea otter is thriving both in and outside the reserve. Ultimately, food limitation and predation by non-human species (particularly bald eagles) will probably limit the growth of the population (ibid.), whereupon maintaining reserves from which human competition for prey is eliminated may help to maintain the species. In the meantime, this charismatic animal is presumably contributing to the retention of a healthily functioning ecosystem in Checleset Bay, both by being there and by acting as a flagship for protection of the reserve.

Cetaceans

Whales and dolphins (cetaceans) have a strong constituency of support as a result of campaigns to protect them and horror over the 'holocaust' they suffered during the whaling era (Whitehead et al., 1999; Lavigne et al., 1999). Mostly for this latter reason, whales are prominent in the COSEWIC lists (see Freedman, this volume) and Canadian actions in establishing marine protected areas have taken particular account of whales.

The ability of cetaceans to communicate intelligently has been popularized, and this has elevated them to beings having 'sentience' (awareness, or feeling). If we consider the denotation of 'sentience' as meaning sensory responsiveness to one's environment, then this is present in all living beings; it is fundamental to life. In fact, there is no clear line between sentient and non-sentient beings. The position of whales in the apparent hierarchy of sentience is simply that whales are more like humans in their awareness than most other animals, and human consciousness has naturally elevated itself in importance as an evolutionary requisite. Cetaceans, it seems, suffer pain and anguish in a similar manner to humans (see Mann, in Twiss and Reeves, 1999). Anguish is much less evident in lobsters and not discernible at all in algae. Why this should be a basis for ranking relative importance for species protection is difficult to rationalize, but it is so in practice because we tend to regard human attributes as specially important.

Prior to 1973 in Canada, whales were actively hunted as a commercial resource and this threatened several populations. Since then, their anthropogenic deaths outside the Arctic have mostly been unintended—through entanglement in fishing gears, ship strikes, pollution, and depletion of their prey by fishing (see Simmonds and Hutchinson, 1996; Whitehead et al., 1999; Twiss and Reeves, 1999). A few entangled cetaceans are still killed by angry Canadian fishermen whose fishing gear has been damaged by them, and others have been shot for stealing fish from nets (see Mulvaney, in Simmonds and Hutchinson, 1996; Northridge and Hofman, 1999), but the numbers are not significant with respect to population status. Some First Nations communities in Canada maintained whale hunts throughout the twentieth century, notably among the Inuit of the North, who traditionally have hunted narwhal, beluga, and bowhead whales particularly. Several populations of beluga whales (*Delphinapterus leucas*) are formally considered to be at risk as a result of their depletion during the commercial whaling era, followed by continuance to the 1990s of traditional hunts at levels that were sustainable in the past but are unsustainable now (Reeves and Mitchell, 1988; Richard, 1990; Doidge and Finley, 1992).

While whale populations within Canada are mostly threatened by activities regarded as routine components of normal human existence, there are more direct threats in some parts of the world (Whitehead et al., 1999; Gambell, 1999). Whaling continues in several forms, notably 'scientific whaling' (conducted within International Whaling Commission guidelines) and 'pirate whaling' from ships registered in nations not represented on the International Whaling Commission (IWC). In addition, conservationists are concerned about the potential for commercial whaling to be disguised as subsistence 'Aboriginal whaling' (Lavigne et al., 1999: 39). Some conservation scientists have called for Canada to rejoin the IWC, which regulates commercial whaling, in order that it might vote for stronger world protection and conservation measures (Whitehead, personal communication). Canada left the IWC in 1981, arguing that it 'no longer has any interest in the whaling industry or in related activities of the IWC' (Department of External Affairs, 1981), but attitudes have changed since then and this position may no longer represent popular Canadian opinion (Lavigne et al., 1999).

Of all marine organisms, whales and dolphins perhaps can benefit least as a group from marine protected areas, simply because they generally have large individual ranges. Furthermore, even when benefit might be obtained, such as in the cases of the northern bottlenose whale, Atlantic right whale, and harbour porpoise (see below), the increased protection that might be afforded by a marine protected area has not been put in place. As Whitehead, Reeves, and Tyack (1999) note, 'For almost all ... marine protected areas, their establishment has caused little or no change in the level of threat faced by cetaceans using the area. Instead, their principal utility has been to forestall future adverse developments'.

Right Whales

The North Atlantic right whale (*Eubalaena glacialis*) and the bowhead whale (*Balaena mysticetus*) are related species of baleen whale that feed on krill. Both are large and relatively slow 'right' whales (i.e., the 'right' whales to hunt because they floated after killing due to the large amount of blubber). They were hunted close to extinction in the past, and both are listed by COSEWIC as endangered (DFO, 1980; Gaskin, 1990).

The bowhead whale is an Arctic species (Mansfield, 1971). It has recently been subjected to a controversial Inuit hunt in Canada, but the take has been very small and essentially symbolic. Major concentrations of bowheads are found seasonally in Lancaster Sound (between Devon Island and Baffin Island) and in Isabella Bay on the northeastern coast of Baffin Island. Both of these areas are relatively productive and support high diversities of Arctic marine wildlife; both have been proposed for protection (Mondor et al., 1995). Community-based management at Isabella Bay is being encouraged within the context of UNESCO Biosphere Reserve designation, an approach recommended by Agardy (1997) for the management of large MPAs. The politics of bowhead whale conservation is unusually complex and strongly affected by the evolution of governance in northern Canada, where the majority of the population is Aboriginal with cultures still founded in hunting.

The Atlantic right whale provides a particularly instructive case of the value of protected areas for whale conservation. The species is critically endangered, and its North Atlantic population has shown virtually no recovery since it was first protected in the 1930s (Gaskin, 1990; Kraus and Brown, 1992; Katona and Kraus, 1999). The species migrates fairly predictably between its wintering grounds off the states of Florida and Georgia and summer feeding grounds in the Gulf of Maine region, notably two areas within Canadian jurisdiction (Kraus and Brown, 1992; Brown et al., 1995). Juveniles congregate in the Roseway Basin on the southern Scotian Shelf. Mothers and calves congregate near Grand Manan Island in the mouth of the Bay of Fundy, and are specially vulnerable there to ship strikes and fishing-gear entanglements. Both areas are marked on Canadian marine charts and 'Notices to Mariners' are provided to ships' captains by DFO when the whales are present. Furthermore, whale researchers and whale watchers keep a close eye on the congregation in the Bay of Fundy, and each individual death or major injury is recorded and reported in news media.

Kraus and Brown (1992) provided a right whale conservation plan for Canada, but it is only partially implemented (Katona and Kraus, 1999). Clearly, more progress is feasible. Nevertheless, recovery of the species will be slow at best because most of the population migrates twice yearly past some of the world's largest ports (Charleston, New York, and Boston) and spends much of its time in intensively exploited fishing grounds. These are extensive threats, and fundamental changes in human attitudes may be needed if these threats are to be significantly lessened. Thus, the best approach for preservation is to reduce the level of threat by a barrage of means and hope that the species will eventually survive its current 'bottleneck' condition.

Harbour Porpoise

The harbour porpoise (*Phocoena phocoena*) is a small threatened cetacean that follows schools of fish, such as herring, on which it preys. Groundfish gill-netting is the major threat to the harbour porpoise in the Gulf of Maine region in both Canada (Gaskin, 1991) and the United States (Northridge and Hofman, 1999). In the late 1980s the Gulf of Maine population was in serious decline, with over 1,800 porpoises dying annually in gill nets on the US side of the line alone (ibid.). The United States Marine Mammal Protection Act requires that mortality approach zero by 2001.

The threat clearly needs to be addressed principally by revising fishing practices. Adding acoustic alarms to bottom-set nets has been shown to deter harbour porpoises from approaching them (Lien et al., 1995), but not with perfect efficiency (Kraus et al., 1997), and thus closure of gill-net fisheries when and where harbour porpoises are present is also an important and feasible protection strategy.

Temporary fishery closures established to protect a 'stock' tend to be regarded as fishery management provisions rather than biodiversity protection provisions, and area-based time-limited fishery closures are therefore generally not considered to be MPAs, even if done to protect a non-target species. In Canada, such closures are put in place using provisions of the Fisheries Act, while permanent MPAs are meant to be established under the Oceans Act. Permanent MPAs are a suitable conservation strategy for some small cetaceans having specific habitat requirements, such as river dolphins (Simmonds and Hutchinson, 1996; Twiss and Reeves, 1999), but permanent MPAs have not been established for the harbour porpoise because of their migratory behaviour and the politics of fishing.

Many marine conservationists would argue that inadequate attention has been paid to the unsuitability of some marine fishery practices and that there must be a fundamental shift in fisheries management towards placing conservation ahead of short-term efficiency of capture. Some gill-net fisheries are indiscriminate and wasteful, having high rates of bycatch as well as creating 'ghost nets' that continue to kill fish when they are lost at sea, as readily happens. Unsustainable practices like these need to be addressed directly, not just within the context of a particular species of concern.

While gill-net fishing practices are the primary issue to be addressed for conservation of the harbour porpoise, additional protection for this species might be obtained within a regional system of MPAs designed for systematic protection of marine bio-diversity, in which the harbour porpoise acted as one of several flagship species.

Beluga Whales of the St Lawrence Estuary

A population of Beluga Whales (*Delphinapterus leucas*) in the St Lawrence estuary has nominally been afforded protection in a marine park maintained by Parks Canada using the provisions of the Parks Act, together with other federal and provincial statutes as required (Dionne, 1995). The remnant whale population, which is listed by COSEWIC as endangered (Pippard, 1983), concentrates at the mouth of the Saguenay River, and the park boundary was selected to capture the region of greatest concentration of the whales (Blane, 1997). The Saguenay Fjord region was first selected as a potential park in 1970 (Dionne, 1995); 30 years later full implementation of the protection originally envisaged still awaits passage of the National Marine Conservation Areas Act. Marine conservation areas are selected to fulfil Parks Canada's intention to represent each of Canada's 29 marine regions in its national system plan, in this case, the St Lawrence estuary.

The current marine park draws tourists to the region, and the prime interest is in whale-watching, for which about 50 boats were operating in 1996 (Blane, 1997). Because the beluga whale is endangered, tour operators are not supposed to focus attention on them, and so the main focus is on fin whales (*Balaenoptera physalus*), which are listed by COSEWIC as vulnerable (Meredith and Campbell, 1987). Blane (1997) reports that although codes of conduct, regulations, fisheries officers, and a local consultative committee are all in place, interference with whale feeding by crowds of whale-watching boats was occurring in 1996, and belugas were being sought as an alternative when too many boats surrounded a pod of fin whales. Rather than providing protection, Blane concludes that 'The creation of the Saguenay-St. Lawrence Marine Park has exacerbated the problem posed by whale-watching by attracting more tourism to this region of the estuary.' Furthermore, she considers that this may get worse in the future, not better, because 'Parks Canada now appears to be giving priority to development [over conservation] by proposing that a purpose statement for MPAs include promotion of MPAs as world class ecotourism destinations (Canadian Heritage, 1997).'

Not only does whale-watching threaten cetaceans within the marine park, but so does pollution. The St Lawrence belugas are the most contaminated whales in the world. They have low birth rates, high rates of infant mortality, and the world's highest rate of death from cancer (Martineau et al., 1994; Blane, 1997). The park authority can do little about pollution because it comes from distant sources associated with two very large rivers, the St Lawrence and the Saguenay. This park, it seems, has little to offer the beluga whales other than a watch over their tragedy.

Northern Bottlenose Whales of the Gully

A more positive example of the application of MPA theory to protection of whales is provided by the Sable Island Gully (usually known simply as 'the Gully') on the Scotian Shelf.

The Gully is a deep submarine canyon at the edge of the continental shelf, roughly 300 kilometres east of Halifax. The canyon was created by an ancient river system that carved its way through a now-eroded and submerged montane system. The resulting modern geomorphological structure has been described as 'unique', in view of its relatively large size, depth, and position. The slope current along the shelf edge enters the canyon and creates a weak gyre, possibly affecting productivity in the canyon. The diversity of fish in the Gully is relatively high (Shackell et al., 1996), as is the whale diversity (Hooker et al., 1999).

An unusual population of northern bottlenose whales (*Hyperoodon ampullatus*) congregates at the mouth of the Gully, where the water is about 500-1,500 metres deep (Reeves et al., 1993; Gowans and Whitehead, 1995). The population is the only known resident population of northern bottlenosed whales, a species that is usually more migratory. The population is listed by COSEWIC as vulnerable (Whitehead et al., 1996). About half the local population of 200–300 animals is present in the Gully at any one time, with the remainder probably foraging along the shelf edge nearby. The Gully region also attracts at least 11 other cetacean species (Gowans and Whitehead, 1995; Hooker et al., 1999).

Largely because of the bottlenose whale population, the Gully was promoted by World Wildlife Fund as an MPA candidate under the provisions of the Oceans Act as soon as this Act was in force. The site had been previously identified as a potential MPA, and voluntary agreements to minimize the number of ships passing through the core area had been made (Faucher and Weilgart, 1992; Faucher and Whitehead, 1995). The major threat to the area comes from nearby oil and gas development, notably from a gas field located on the very edge of the canyon. In 1998, DFO conducted an intensive review of scientific studies of the Gully (Harrison and Fenton, 1998), and later that year the Minister of Fisheries and Oceans announced that it was to be a 'pilot MPA' (now described as an 'Area of Interest').

The amount of protection actually afforded in the Gully pilot MPA is debatable (see Hooker et al., 1999). Ships supplying the gas-well drilling operations avoid the area, as per the voluntary agreement that pre-existed the announcement of the pilot MPA. A proposal to lease the Gully for oil and gas exploration was withdrawn by the Canada-Nova Scotia Offshore Petroleum Board in 1999 following complaint, and at least one fishing fleet voluntarily excluded the area from its zone of operation in 1999 (D. Fenton, personal communication). On the other hand, a northern bottlenose whale has been reported to have been found entangled in surface long-line fishing gear in the Gully in 2000. The fishing gear had probably been set to catch swordfish. The swordfish long-line fishery is notorious for its wasteful bycatch, and its place in an MPA is therefore questionable.

There has been almost no progress in drawing up a management plan for the Gully, in part because there is continued opposition from fishing interests and also because DFO intends to create its management plan within the context of a plan for resource management on the eastern Scotian Shelf as a whole. Whether this is a politically wise and socially conscious approach to integrated marine management, or a strategy for holding marine biodiversity in the Gully hostage to commercial interests, is bound to be debated.

Examples of Unlisted Species at Risk

A comprehensive review of scientific opinion concerning which species might be brought forward for listing as being 'at risk' would probably produce a huge list of marine species about which too little is known, as well as a few that deserve immediate attention. I have selected three cases that together tell an interesting story of contrasts. The huge barndoor skate declined to near extinction without anyone noticing, while the Atlantic salmon has been a subject of intense attention and numerous conservation efforts for more than a century. Canada's deep-sea corals are the least studied and least understood of these three, yet as a group they are clearly the most fundamental to marine ecosystem health.

Barndoor Skate

Casey and Myers (1999) examined long-term databases for fish species surveyed using experimental trawl tows in the northwest Atlantic. They found that the largest skate in the region had declined drastically throughout the range examined. There is no fishery directed at the barndoor skate (*Raja laevis*); it is an incidental bycatch in the mobile-gear (bottom trawl) fishery. In the northern part of the area examined, no individuals were caught during the most recent 20 years, but it was previously a regular bycatch. Casey and Myers (1999) estimate that the population has declined to less than 0.1 per cent of previous abundance. The large size and relatively late sexual maturity of this species (11 years) makes it specially vulnerable because it is readily caught before it is able to reproduce.

Casey and Myers explain that 'Failure to examine historical data has resulted in the largest skate in the northwest Atlantic being driven to near extinction without anyone noticing.' They also state: 'Perhaps the only hope for the long-term survival of this species is to designate an area protected from trawling on all the banks that is sufficiently large to allow for a self-sustaining population. A protected area would also provide a simple and effective means to conserve other species.' Like the sea otter and the harbour porpoise, perhaps the barndoor skate can become a flagship for other less obvious marine species.

Atlantic Salmon

While the decline of the barndoor skate went unnoticed at first, this was not so for the Atlantic salmon (*Salmo salar*). This fish is prized by anglers for its feisty spirit, loved by gourmets for its rich taste, and admired by naturalists for its spectacular ability to leap over rapids and waterfalls. Its progressive decline in various parts of its North Atlantic range has been noted for decades (see, e.g., Baum 1997).

The Atlantic salmon is an anadromous species, that is, it lives some of its life in the marine environment but enters rivers to spawn. Each river has a distinct population that passes spawning fidelity from generation to generation. In Canada, the wild species is limited to rivers that drain into the Atlantic Ocean, and in the southern part of the range the decline has been most severe. Remarkably, the salmon is not listed as being at risk either by COSEWIC or under the US Endangered Species Act, despite the fact that some rivers in Maine and Nova Scotia that once were rich with salmon now have none at all. Many distinct populations are extinct.

The threats to Atlantic salmon are relatively well understood and have been clearly set out in a recent review by the US Anadromous Atlantic Salmon Biological Review Team (1999). The major threats of historical origin are: river barrages, spawning habitat destruction, reduction in population sizes by overfishing, and reduction in genetic fitness due to poorly planned stocking programs. The major additional current threats are: water removal from rivers for agriculture, sediments and pesticides entering rivers, diseases and parasites introduced via salmon aquaculture, genetic intrusion by farmed salmon, recreational fishing in rivers and commercial fisheries at sea, and elevated water temperatures in spawning streams. These threats are cumulative, and represent a potent brew. Under normal conditions, natural predation by fish, birds, and mammals would not be considered a 'threat' but would be regarded as part of the wholeness that maintains genetic health through natural selection. Whether this holds true now is a matter of debate.

The American review team regards the remnant wild salmon populations of eight rivers in Maine as a 'distinct population segment' (DPS), and they consider that this DPS is 'in danger of extinction'. Under normal circumstances, this would provide a strong rationale for listing the species under the US Endangered Species Act, but the circumstances are not normal. Earlier attempts to list the species produced strong reaction from lobbyists and politicians in Maine. They fear that listing will hurt the sport fishing industry, the aquaculture industry, and perhaps affect farming. As a result, the state of Maine has vigorously opposed listing and has developed salmon conservation programs, including the use of funding and advice from the federal fisheries service. Bauer (1997) argues that this response may be as effective as any response that might be triggered by listing, and therefore that listing itself is not the only value of an Endangered Species Act; the fear of listing can also be effective in increasing the protection afforded a species.

The situation in Canada is different. Atlantic salmon river populations in the northern part of the range (Newfoundland) are healthy, while those in the south, particularly in southern Nova Scotia, are extinct or in serious jeopardy. Acidification of rivers and streams by acid rain is the major culprit in southern Nova Scotia, for other risk factors are not elevated. Storms gather sulphur dioxide as they track over industrial regions to the southwest, depositing much of this in Maine and southern Nova Scotia. The granite bedrock of southern Nova Scotia provides almost no chemical buffering capacity. Salmon populations have been most resilient in rivers having drainage basins that include limestone bedrock, and river liming programs have sometimes proved beneficial in the short term.

In the upper Bay of Fundy, between Nova Scotia and New Brunswick, there is a special salmon group. Unlike most Atlantic salmon, they do not migrate into the open ocean but remain in the bay, migrating between there and their specific rivers. This behaviour is unique to the group. Several of these rivers have lost their populations completely, and others retain very small remnants (Hutchings, personal communication). This group deserves special conservation attention and appears to be a good candidate for listing by COSEWIC. Protected areas around the mouths of Bay of Fundy rivers having remnant Atlantic salmon populations is a feasible element of a comprehensive conservation strategy. In such areas, any activity that might trap individuals or interfere with salmon migration would not be permitted, but other activities could continue, in a manner similar to sanctuaries set aside for migrating birds.

It is instructive to wonder why the Atlantic salmon was not listed as vulnerable or threatened years ago in Canada. The answer almost certainly lies in the politics of the situation. Anglers, who are intensely aware of the situation, are caught between their love of the fish and their love of fishing. If the salmon is formally acknowledged as endangered, will this mean that anglers cannot even practice 'catch-and-release', their favoured option? If an anadromous species is listed as endangered, does this mean that the same protection strategies must be afforded in rivers, bays, and open ocean? If a marine fisherman accidentally caught one, would the penalty be the same as for a poacher angling on a river? Would the whole of the Bay of Fundy have to become a protected area in order to protect its special population? Atlantic salmon have been purposefully introduced to the wild in some faraway places, such as New Zealand, and it is a species of choice in aquaculture in many parts of the world. The species is more plentiful (but less genetically diverse) in captivity than it is in the wild. How does this affect its conservation status elsewhere? By not listing the Atlantic salmon, finding answers to vexing questions like these can be avoided, but we can expect this particular debate to heat up when species-at-risk legislation is finally introduced and passed in Parliament.

Deep-Sea Corals

As the discussion to this point will have made clear, the belief that marine natural systems are intrinsically resilient has led to insufficient awareness that marine species may be at risk of extinction. Furthermore, marine species at risk are most likely to be found at the bottom of the ocean and are therefore obscure. In addition, with the exception of mammals and birds, marine species do not have a large popular constituency of support and tend to be regarded as important only if they have utilitarian value. Marine species are also more likely to be discounted for being 'slimy' or 'ugly'.

An interesting exception to this general rule is the group of colonial benthic anthozoans (sea anemones and their relatives) known as corals. Popularization of tropical coral reefs through underwater photography and tourism has produced a strong constituency of support for the spectacular colourful ecosystems founded upon the biophysical structures created by coral organisms. There is now worldwide concern about the status of coral reef systems for many reasons (Wilkinson, 1996).

Canadian corals are azooxanthellate, that is, they do not have photosynthetic organisms associated with them, unlike many of their shallow-water tropical zooxanthellate cousins. They mostly live in aphotic (dark) ocean zones, at depths greater than 100 metres, and are commonly called deep-sea corals, although other names are also current. At these depths, horny (gorgonacean) corals form 'forests' wherever there is suitable hard substrate for settlement and a sufficient steady current to provide a supply of organic particles that the coral polyps filter from the passing water. The 'trees', a term commonly used by fishermen to describe gorgonian coral colonies (see Breeze et al., 1997), create underwater forests of 1–3 metres in height.

In the North Sea, mounds created by the stony (scleractinian) coral *Lophelia pertusa* may be very large, covering several square kilometres of ocean floor and rising thirty or more metres from the bottom (Wilson, 1979; Jensen and Frederiksen, 1992; Roberts, 1997). Specimens of the calcareous skeletons of *Lophelia pertusa* have been collected off Nova Scotia, but it is not known whether living colonies still exist there (Breeze et al., 1997; Breeze and Davis, 1998).

Like their well-known tropical counterparts, coral-based ecosystems of temperate marine regions have high species richness (Jensen and Frederiksen, 1992). Temperate deep-sea coral ecosystems are too poorly known for it to be possible to state that they create essential habitat for associated species, but given the similarity of the ecosystems to tropical coral reefs, it would be precautionary to assume that they are just as critical and just as fragile (Roberts, 1997).

Deep-sea corals are not generally regarded as a resource, and they have not been collected in significant quantities for commercial purposes in Canada. They are at substantial risk, however, and have been knocked down in large quantities by fishing activity, mostly by mobile gears. In Canada, the most injurious fishing method is called 'dragging', and the most common activity in coral-rich areas is 'groundfish dragging' using otter trawls, often with 'rock-hopper gear'. In this method, a voluminous net is dragged along the ocean bottom, catching fish as it goes, while also knocking down anything that stands in the way. Groundfish long-line fishing (which is quite different from surface long-lining) also damages corals, but poses a significantly lesser threat within coral areas. In this method, the fishing gear consists of a line of baited hooks anchored to the bottom and then collected some time later. During the recovery of the fishing line, corals may get snagged or knocked over.

There have been reports that deep-sea corals have been purposefully cleared out of the way in large quantities by some dragger fishermen, who regard them as a nuisance (Breeze et al., 1997). Other Canadian dragger fishermen report that they try to stay away from corals because their nets get filled with them when they stray into coral-rich areas. Long-line fishermen, on the other hand, tend to regard deep-sea corals as important habitat for the fish species they seek to catch, and so they sometimes try to fish among corals while damaging them as little as possible (Sanford Atwood, fisherman, personal communication).

No true MPAs have yet been set aside in Canada to protect coral ecosystems, but official reports supporting at least two 'areas of interest' (also known as 'pilot marine protected areas'), identified under the Oceans Act process for selecting potential marine protected areas, have included mention of the presence of corals. These are the Sable Island Gully and Race Rocks (near Victoria, Vancouver Island) (DFO, 2000). Restrictions on activities in these regions are voluntary as long the marine protected area is described as a 'pilot', however.

A very interesting proposal for deep-sea coral protection is an area proposed by Derek Jones and Sanford Atwood of the Canadian Ocean Habitat Protection Society (COHPS). This non-profit group is based in Newellton, Cape Sable Island, at the southernmost tip of Nova Scotia. The communities of Cape Sable Island have been steeped in the sea for generations; they owe their existence entirely to the rich fishing grounds of Georges Bank and Browns Bank, to which they are the closest landfall. Georges Bank lies at the outer edge of the Gulf of Maine, and Browns Bank is the most western of the Scotian Shelf banks. The founders of COHPS are hook-and-line fishermen with a unique aggregate of experience for a conservation group. Their fathers and grandfathers fished for cod, haddock, and halibut, and they have done the same, trained from boyhood. The group came into existence as a direct result of the failure of local fisheries and the belief of these fishermen that habitat destruction has been an important contributing factor.

According to the local knowledge accumulated by COHPS, the richest remnant of deep-sea corals on Canada's east coast lies at the mouth of the Northeast Channel, between Browns and Georges Banks, just to the south of Browns Bank. The area is subject to complex currents, varying with depth and tide, which make fishing diffi-cult but are excellent for corals. Sediment is swept from hard substrate, and there is a steady supply of nutrients for these filter-feeding organisms. The common 'tree' species *Paragorgia arborea* and *Primnoa resedaeformis* are plentiful, and several less common species are present as well.

COHPS identified the area of interest with care so as to maximize coral protection while minimizing interference with traditional patterns of fishing. Following the identification of co-ordinates by COHPS in 1999, the area was surveyed by the Canadian Hydrographic Service using multi-beam sonic bathymetry, and thus a good picture of the bottom topography is now available. A submarine survey should be conducted next so that an MPA can be delineated for the purpose of protecting a wild coral forest on Canada's east coast. Extensive surveying of Canada's Atlantic and Pacific coastal shelves for coral remnants is also wise, in order to assess the damage done and restore as much as is feasible.

SYSTEMATIC PLANNING FOR MARINE PROTECTED AREAS

Systematic selection of MPAs towards the creation of an integrated network is gener-ally recognized as a rational and precautionary approach to the conservation of marine biodiversity (Ballantine, 1995; Dayton et al., 1995). A substantial body of scien-tific evidence indicates that such an approach may also protect and enhance fisheries (Bohnsack, 1993; Davis, 1998; Johnson et al., 1999; Dayton et al., 2000). Despite this, progress towards creating systems of MPAs has been extremely slow, as clearly illus-trated by the lack of real progress by Parks Canada in establishing an effective system of marine parks, despite at least 30 years of work to this end. One is bound to wonder why such a simple and good idea has not been enthusiastically implemented in a wealthy nation like Canada.

Among the reasons for this slow progress is surely that the ocean realm is generally regarded as a source of resources to be exploited by anyone brave enough to venture out on the seas. There is a historically well-earned tradition of respect for the bravery of sailors, and with this has come a sense that those who venture onto the seas own them. The creation of a system of MPAs tends to be an office enterprise, based on analysing scientific reports and consulting with interested parties who do not necessarily go to sea in boats. This clash of life experiences has inevitably led to conflict, and it has therefore become politically inexpedient to press on with conservation plans that relate to protecting marine biodiversity rather than marine resources. Public support for the protection of visible beauty on land has not yet extended to the invisible depths of the ocean. The protection of endangered species has been one of few rationales for establishing MPAs that has successfuly climbed over this wall of conflict. Unfortunately, it is a rationale of last resort, and a poor one in ideal terms. The best rationale for MPAs is precautionary, not reactionary. Unfortunately, we have waited until extinction has approached crisis proportions in the marine realm before acting, and thus a calm precautionary approach to MPA systems planning is no longer possible if marine biodiversity is actually to be protected. For this reason, catalysts for action have been sought.

Many of the pilot MPAs recently selected in Canada make sense as catalysts. The Gully is a spectacular canyon, home to a threatened whale population; the Musquash Estuary in New Brunswick is the last estuary feeding into the Bay of Fundy to remain largely unaffected by development; the algae-covered Bowie Seamount near the Queen Charlotte Islands rises to within 20 metres of the surface from an abyssal depth of 3,000 metres; the Endeavour Hot Vents are over 2000 metres below the surface of the Pacific Ocean, but have a rich diversity of amazing life forms; the Race Rocks near Victoria, BC, have a rich diversity of marine life forms living close enough to the surface that divers can marvel at them. These pilot MPAs are special places (for detail, see DFO, 2000).

Special nature reserves will not protect elements of ordinary marine biodiversity. Once these catalysts have caught the public imagination, Canada will have to create a system of MPAs that, like the national parks system, protects representative elements of ordinary ocean wilderness (Davis, 1998). To achieve this, cornerstones for the system will be needed, such as the ocean wilderness proposed for the Hague Line separating Canadian and US waters in the Gulf of Maine (Willison, 1997), which may act both to preserve marine wilderness and protect fisheries (McGarvey and Willison, 1995).

REFERENCES

Agardy, M.T. 1997. *Marine Protected Area and Ocean Conservation*. San Diego: Academic Press.

Anadromous Atlantic Salmon Biological Review Team. 1999. *Review of the Status of Anadromous Atlantic Salmon (Salmo salar) under the U.S. Endangered Species Act*. Washington: US Fish and Wildlife Service.

Ballantine, W.J. 1991. *Marine Reserves for New Zealand.* Leigh Laboratory Bulletin 25, University of Auckland.

——. 1995. 'Networks of "no-take" marine reserves are practical and necessary', in Shackell and Willison (1995: 13–20).

Barr, B.W. 1995. 'The U.S. National Marine Sanctuary Program and its role in preserving sustainable fisheries', in Shackell and Willison (1995: 165–73).

Baum, E. 1997. *Maine Atlantic Salmon, a National Treasure.* Hermon, Me: Atlantic Salmon Unlimited.

Blane, J.M. 1997. *Marine Protected Areas as a Tool for Cetacean Conservation.* Project Report, Marine Affairs Program, Dalhousie University.

Bohnsack, J.A. 1993. 'Marine reserves: they enhance fisheries, reduce conflicts, and protect resources', *Oceanus* 38, 3: 63–71.

Breeze, H., and D.S. Davis. 1998. 'Deep sea corals', in Harrison and Fenton (1998: 113–20).

——, ——, and M. Butler. 1997. *Distribution and Status of Deep Sea Corals off Nova Scotia.* Marine Issues Committee Special Publication #1. Halifax: Ecology Action Centre.

Brown, M.W., J.M. Allen, and S.D. Kraus. 1995. 'The designation of seasonal Right Whale conservation areas in the waters of Atlantic Canada', in Shackell and Willison (1995: 90–8).

Canadian Heritage, Parks Canada. 1997. 'Charting the course: towards a Marine Conservation Areas Act', discussion paper. Ottawa: Minister of Public Works and Government Services.

Carlton, J.T., J.B. Geller, M.L. Reaka-Kudla, and E.A. Norse. 1999. 'Historical extinctions in the sea', *Annual Review of Ecology and Systematics* 30: 515–38.

Casey, J.M., and R.A. Myers. 1998. 'Near extinction of a large, widely distributed fish', *Science* 281: 690–2.

Davis, G.E. 1998. 'What good is marine wilderness?', in N.W.P. Munro and J.H.M. Willison, eds, *Linking Protected Areas with Working Landscapes.* Wolfville, NS: SAMPAA, 133–7.

Dayton, P.K., S.F Thrush, M.T. Agardy, and R.J. Hofman. 1995. 'Environmental effects of marine fishing', *Aquatic Conservation: Marine and Freshwater Ecosystems* 5: 205–32.

——, E. Sala, M.J. Tegner, and S. Thrush. 2000. 'Marine reserves: parks baselines and fishery enhancement', *Bulletin of Marine Science.*

Department of External Affairs. 1981. 'Canada withdraws from the International Whaling Convention and Commission', press release, Government of Canada.

DFO (Department of Fisheries and Oceans). 1980. 'Status report on the bowhead whale *Balaena mysticetus*', Ottawa: COSEWIC.

——. 2000. Web site: www.pac.dfo-mpo.gc.ca/oceans/mpa

Dionne, S. 1995. 'Creating the Saguenay Marine Park—a case study', in Shackell and Willison (1995: 189–96).

Doidge, D.W., and D. Finley. 1992. 'Status report on the Beluga (White Whale) (Eastern High Arctic/Baffin Bay population) *Delphinapterus leucas* in Canada'. Ottawa: COSEWIC.

Edge, T.A. 1984. 'Status report on the Atlantic Whitefish *Coregonus huntsmani*'. Ottawa: COSEWIC.

Faucher, A., and L.S. Weilgart. 1992. 'Critical marine habitat in offshore waters: the need for protection', in Willison et al. (1992: 75–8).

—— and H. Whitehead. 1995. 'Importance of habitat protection for the Northern Bottlenose Whale in the Gully, Nova Scotia', in Shackell and Willison (1995: 99–102).

Gambell, R. 1999. 'The International Whaling Commission and the contemporary whaling debate', in Twiss and Reeves (1999: 179–98).

Gaskin, D.E. 1990. 'Updated status report on the Right Whale *Eubalaena glacialis* in Canada'. Ottawa: COSEWIC.

——. 1991. 'Status of the Harbour Porpoise *Phocoena phocoena* in Canada'. Ottawa: COSEWIC.

Gowans, S., and H. Whitehead. 1995. 'Distribution and habitat partitioning by small odontocetes in the Gully, a submarine canyon on the Scotian Shelf', *Canadian Journal of Zoology* 73: 1599–1608.

Gubbay, S., ed. 1995. *Marine Protected Areas: Principles and Techniques for Management*. London: Chapman and Hall.

Harrison, G., and D.G. Fenton, eds. 1998. *The Gully: A Scientific Review of its Environment and Ecosystem*. Canadian Stock Assessment Secretariat Research Document 98/83. Ottawa: DFO.

Hooker, S.K., H. Whitehead, and S. Gowans. 1999. 'Marine protected area design and the spatial and temporal distribution of cetaceans in a submarine canyon', *Conservation Biology* 13: 592–602.

Houston, J., and D.E. McAllister. 1989. 'Status report on the Bering Wolffish *Anarhichas orientalis*'. Ottawa: COSEWIC.

Jensen, A., and R. Frederiksen. 1992. 'The fauna associated with the bank-forming deepwater coral *Lophelia pertusa* (Scleractinia) on the Faroe Shelf', *Sarsia* 77: 53–69.

Johnson, D.R., N.A. Funicelli, and J.A. Bohnsack. 1999. 'Effectiveness of an existing estuarine no-take fish sanctuary within the Kennedy Space Center, Florida', *North American Journal of Fisheries Management* 19: 436–53.

Katona, S.K., and S.D. Kraus. 1999. 'Efforts to conserve the North Atlantic Right Whale', in Twiss and Reeves (1999: 311–31).

Kelleher, G., and R. Kenchington. 1992. *Guidelines for Establishing Marine Protected Areas*. Gland, Switz.: IUCN.

——, C. Bleakley, and S. Wells. 1995. *A Global Representative System of Marine Protected Areas*, 4 vols. Washington: The World Bank.

Kenyon, K.W. 1969. 'The sea otter in the eastern Pacific Ocean', *North American Fauna* 68: 1–352.

Kirk, D.A. 1985. 'Status report on the Labrador Duck *Camptorhynchus labradorius*'. Ottawa: COSEWIC.

Koslow, J.A., and K. Gowlett-Holmes. 1998. *The Seamount Fauna off Southern Tasmania: Benthic Communities, Their Conservation and Impact of Trawling*. Report on FRDC Project 95/058. Environment Australia.

Kraus, S.D., and M.W. Brown. 1992. 'A right whale conservation plan for the waters of Atlantic Canada', in Willison et al. (1992: 79–85).

——, A.J. Read, A. Solow, K. Baldwin, T. Spradlin, E. Anderson, and J. Williamson. 1997. 'Acoustic alarms reduce porpoise mortality', *Nature* 388: 525.

Lavigne, D.M., V.B. Scheffer, and S.R. Kellert. 1999. 'The evolution of North American attitudes toward marine mammals', in Twiss and Reeves (1999: 48–86).

Lien, J., et al. 1995. 'Field tests of acoustic devices on groundfish gillnets: assessment of effectiveness in reducing harbour porpoise by-catch', in R.A. Kastelein et al., eds, *Sensory Systems of Aquatic Mammals*. Woerden, The Netherlands: De Spil, 349–64.

Mansfield, A.W. 1971. 'Occurrence of the Bowhead or Greenland Right Whale (*Balaena mysticetus*) in Canadian Arctic Waters', *Journal of the Fisheries Research Board of Canada* 28: 1873–5.

Martineau, D., et al. 1994. 'Pathology and toxicology of beluga whales from the St. Lawrence Estuary, Quebec, Canada: Past, present and future', *The Science of the Total Environment* 154: 201–15.

McGarvey, R., and J.H.M. Willison. 1995. 'Rationale for a marine protected area along the international boundary between U.S. and Canadian waters in the Gulf of Maine', in Shackell and Willison (1995: 74–81).

Meredith, G.N., and R.R. Campbell. 1987. 'Status report on the Fin Whale *Balaenoptera physalus* in Canada'. Ottawa: COSEWIC.

Mondor, C., F. Mercier, M. Croom, and R. Wolotira. 1995. 'Marine Region 4, Northwest Atlantic', in Kelleher, Bleakley, and Wells (1995, vol. 1: 105–26).

Montevecchi, W.A., et al. 1995. 'National recovery plan for the Harlequin Duck in eastern North America'. Report no. 12. Ottawa: Recovery of Nationally Endangered Wildlife Committee.

Murray, S.N., et al. 1999. 'No-take reserve networks: sustaining fishery population and marine ecosystems', *Fisheries* 24, 11: 11–25.

Northridge, S.P., and R.J. Hofman. 1999. 'Marine mammal interactions with fisheries', in Twiss and Reeves (1999: 99–119).

Ottesen, P., and R. Kenchington. 1995. 'Marine conservation and protected areas in Australia: what is the future?', in Shackell and Willison (1995: 151–64).

Parker, B. 1988. 'Status report on the Deepwater Cisco *Coregonus johannae*'. Ottawa: COSEWIC.

Parks Canada. 1994. 'National Marine Conservation Areas Policy', in *Guiding Principles and Operational Policies*. Ottawa: Supply and Services Canada, 43–61.

——. 1995. *Sea to Sea to Sea: Canada's National Marine Conservation Areas System Plan*. Ottawa: Supply and Services Canada.

Pauly, D., and V. Christensen. 1995. 'Primary production required to sustain global fisheries', *Nature* 374: 255-7.

——, ——, J. Dalsgaard, R. Froese, and F. Torres. 1998. 'Fishing down the marine food webs', *Science* 279: 860–3.

Pippard, L. 1983. 'Status report on the beluga (White Whale) (St. Lawrence River population) *Delphinapterus leucas* in Canada'. Ottawa: COSEWIC.

Reeves, R.R., and E. Mitchell, 1988. 'Status report on the Beluga (White Whale) (Ungava Bay and Eastern Hudson Bay populations) *Delphinapterus leucas* in Canada'. Ottawa: COSEWIC.

——, ——, and H. Whitehead. 1993. 'Status of the northern bottlenose whale, *Hyperoodon ampullatus*', *Canadian Field-Naturalist* 107: 490–508.

Richard, P.R. 1990. 'Status report on the Beluga (White Whale) (S.E. Baffin Island/Cumberland Sound population) *Delphinapterus leucas* in Canada'. Ottawa: COSEWIC.

Roberts, C.M., and J.P. Hawkins. 1999. 'Extinction risk in the sea', *Trends in Ecology and Evolution* 14: 241–6.

Roberts, M. 1997. 'Coral in deep water', *New Scientist* 155: 40–3.

Rodway, M.S. 1990. 'Status report on the Marbled Murrelet *Brachyramphus marmoratus*'. Ottawa: COSEWIC.

Safina, C. 1998. 'Scorched-earth fishing', *Issues in Science and Technology* 14: 33–6.

Shackell, N.L., and J.H.M. Willison, eds. 1995. *Marine Protected Areas and Sustainable Fisheries.* Wolfville, NS: SAMPAA.

——, P. Simard, and M. Butler. 1996. 'Potential protected area in The Gully region, Scotian shelf'. Report to World Wildlife Fund Canada.

Simmonds, M.P., and J.D. Hutchinson, eds. 1996. *The Conservation of Whales and Dolphins: Science and Practice.* Chichester, UK: John Wiley and Sons.

Soulé, M.E. 1991. 'Conservation: tactics for a constant crisis', *Science* 253: 744–50.

Twiss, J.R., Jr, and R.R. Reeves, eds. 1999. *Conservation and Management of Marine Mammals.* Washington: Smithsonian.

Watling, L., and E. Norse. 1998. 'Disturbance of the seabed by mobile fishing gear: a comparison to forest clearcutting', *Conservation Biology* 12: 1180–97.

Watson, J.C., G.M. Ellis, T.G. Smith, and J.K.B. Ford. 1996. 'Updated status of the sea otter, *Enhydra lutris*, in Canada'. Ottawa: COSEWIC.

Whitehead, H., A. Faucher, S. Gowans, and S. McCarrey. 1996. 'Status of the Northern bottlenose Whale (Gully population), *Hyperoodon ampullatus*, in Canada'. Ottawa: COSEWIC.

——, R. Reeves, and P. Tyack. 1999. 'Science and the Conservation, Protection and Management of Wild Cetaceans', in J. Mann, R.C. Connor, P.L. Tyack, and H. Whitehead, eds, *Cetacean Societies.* Chicago: University of Chicago Press, 308–32.

Wilkinson, C.R. 1996. 'Global change and coral reefs—impacts on reefs, economies and human cultures', *Global Change Biology* 2: 547–58.

Willison, J.H.M. 1997. 'Calming troubled waters: an international peace park in the Gulf of Maine', *Northern Forest Forum* 5, 6: 20–1.

—— et al., eds. 1992. *Science and the Management of Protected Areas.* Amsterdam: Elsevier.

Wilson, J.B. 1979. ' "Patch" development of the deep-water coral *Lophelia pertusa* (L.) on Rockall Bank', *Journal of the Marine Biological Association of the United Kingdom* 59: 165–77.

PART TWO

FROM SCIENCE
TO POLICY

The Politics of Endangered Species: A Historical Perspective

STEPHEN BOCKING

In August 1995 the federal government released its legislative proposal for a Canadian Endangered Species Act. Since then, federal initiatives, including tabling of a revised Species at Risk Act in April 2000 (and its demise in October 2000 with the federal election call), have elicited reactions from many parties. Provincial governments have stressed that wildlife remains under their jurisdiction. Environ-mental groups have challenged the government to act more aggressively, while industry groups have expressed concern about the cost of protecting species and habitats. Landowners have stressed the need to compensate those affected when protected species are found on private property. Scientists have described a global crisis of extinction, and have urged that science, not politics, determine which species are to be protected.

These various attitudes highlight several aspects of the politics of endangered species. The federal initiative reflects the need for a national approach: species do not recognize provincial borders; neither can their protection. Provincial views epitomize the challenges of national initiatives in a decentralized federation. The diverse voices of citizens' groups illustrate how concerns about endangered species have broadened far beyond a few specialist and interest groups. Opposition from industry and some landowners reflects how the controversy turns on different ways of valuing the land and wildlife; some of these ways have been part of Canadian society for centuries, while others have developed more recently. Participation by scientists continues their long-standing efforts to define the issue of endangered species, and exemplifies the ongoing demand for scientific information to guide efforts to protect species.

These aspects can also be traced back through several decades of events. For example, the creation of the Committee on the Status of Endangered Wildlife in Canada (COSEWIC) in 1977 was the result of activism by environmentalists, scientists, and others who saw the creation of a single national list of endangered species as essential to a coherent, science-based program of species protection. These activists, of course, also had to be cognizant of provincial concerns regarding federal infringement on their jurisdiction.

But to understand why endangered species have become a political issue, these events must be interpreted in the context of trends in the recent history of Canadian politics and society. These include:

- the expansion and evolution of resource exploitation and consumption, with a multitude of impacts on wildlife;
- increased scientific study of the environment, leading to greater understanding of the ecology of endangered species, as well as an increasing tendency to define issues in scientific terms;
- an increasingly pluralistic politics of the environment, as an ever-widening array of agencies and groups have begun to participate in debates and initiatives;
- the emergence of new environmental values among Canadians.

The growing public awareness of the environment and new environmental values are the result, in part, of a decreasing emphasis on material production in favour of a stronger focus on the enjoyment of amenities—intact wilderness, clean air and water—only obtainable through a protected environment.

A historical perspective can contribute to our understanding of contemporary endangered species politics. By placing the trends, actions, and events of recent years within a broader context we can learn how and why they emerged within a changing economy and society, shaped by new collective values and knowledge. We can begin to understand the changes that have occurred in environmental politics by returning briefly to a time of different values and attitudes.

AN ABUNDANCE OF GAME

In 1942 wildlife managers and scientists travelled from across the continent to Toronto for their annual North American Wildlife Conference. Provincial officials attending told their colleagues of the bounties of the Canadian wilderness. 'We have the finest big game country in North America; more big game than we know what to do with', bragged J.G. Cunningham, the BC Game Commissioner. E.S. Huestis, Fish and Game Commissioner for Alberta, echoed Cunningham's boast, and stressed the benefits of the province's parks: 'We have one of the greatest natural game preserves in the world, the national parks. They grow our big game for us on the eastern side of the Rockies, on the eastern slope, and what they can't feed roam over into our territory, and we can supply you with some very fine shooting.' Their fellow game commissioners continued these themes, stressing abundant wildlife, opportunities for hunters, and game officials' aim of promoting its full use.

However, Charles Frémont, superintendent of the Fish and Game Department of Quebec, had different points to make. Among his province's attractions, he noted, were huge colonies of puffins, 'not interesting on the table, but very, very interesting when you look up their habits and their peculiar system of doing their fishing, raising their young, breeding, and so forth.' He also noted Quebec's success in protecting snow geese, which had increased from 4,000 to 22,000 over the last 30 years (North

American Wildlife Conference, 1942). Protection of wildlife and their enjoyment by means other than hunting were his chosen themes.

Celebration of the abundance of wildlife and the opportunities they provided for hunting and trapping, together with concern for the status of a few species, defined much of the politics of wildlife in Canada prior to 1960. These themes were also exemplified in the work of the Canadian Wildlife Service (CWS). It focused, in part, on species important to hunters and trappers, particularly in northern Native communities: the caribou, Arctic fox, and other game. Its most significant area of activity, however, stemmed from its responsibility for administering the 1916 Migratory Birds Convention Treaty.

There were a few concerns regarding endangered species—the passenger pigeon and a few other extinct species still cast a shadow. When CWS officials drew attention in the 1950s to a northern 'caribou crisis' caused by excessive hunting, the prospect of another episode like the near-extinction of the bison was likely on their minds (Banfield, 1956; Tener, 1960). Much attention also focused on a few highly prominent species of birds. The trumpeter swan was thought to be approaching extinction, with an estimated population in 1916 of as few as 100, from which it had since recovered to possibly 900 by 1947 (Munro, 1962; Lewis, 1947). In 1950 Ron Mackay was assigned by the CWS to determine how many still existed and where they were located. CWS efforts to restore this species included captive breeding and release programs and the transfer of family groups to new habitats (Burnett, 1999). The whooping crane, however, became the most prominent symbol of early CWS efforts regarding endangered species. By the late 1940s the species had declined to as few as 21 individuals. Once its nesting areas were discovered in Wood Buffalo National Park in 1954, the CWS became involved in long-term studies and recovery efforts, in co-operation with American authorities (Burnett, 1999: 141–3).

Wildlife politics in this era had few players. The CWS was dominant. While provincial agencies for game or natural resources carried out certain functions, such as the regulation and promotion of hunting and fishing, the CWS conducted the largest amount of wildlife research and effectively defined through its activities the state of wildlife management across the country. So weak were provincial agencies (with the exception of Ontario) that they would periodically call for greater federal involvement in wildlife activities—in striking contrast to a later era when provinces would defend their jurisdiction (see, e.g., Department of Northern Affairs and National Resources, 1956: 30–1). Neither had universities or private organizations developed significant capabilities in wildlife science.

Canadians' economic relations with the wilderness were still substantially defined in terms of consumption: meat and fur from wildlife; wood, water, and minerals from the rest of the frontier. While, famously, Banff National Park was created to cash in on its tourism potential, and wilderness areas elsewhere attracted camps and lodges, before 1960 the call of the wild lured only a more mobile minority (Page, 2000). Until 1954 eight provincial parks were considered sufficient for all of Ontario, and highways were not yet available to convey tourists into the Rocky Mountains. Neither the infrastructure nor the economic prosperity yet existed to enable outdoor recreation on a mass scale.

Diverse Impacts, Actors, and Species

By the early 1960s, however, new ideas about species were developing. Most crucial was the idea that endangered species had to be defined more broadly, to encompass a wider range of species as well as a wider variety of threats. Extinction concerns could no longer be defined only in terms of a few highly visible species, such as the bison or whooping crane. And not just bullets, but more subtle, more pervasive impacts would be viewed as threats, such as the loss of wetlands and other habitats, and pesticides and other toxic chemicals.

Scientists played a crucial role in developing these new views. Within the CWS, the national museums, universities, and other institutions, scientists studied a wider range of species, albeit while still concentrating on birds and mammals. For the first time sustained research began on predators, including wolves, coyotes, and polar bears, while seabirds, shorebirds, and songbirds attracted CWS researchers, reflecting a broadening of that agency's mandate to include non-game birds (Pimlott, 1961; Harington, 1964; Erskine, 1977). By 1970 enough information had accumulated to provide the basis for a first list of 92 rare and endangered species, reviewed in a set of articles in the *Canadian Field-Naturalist* (McAllister et al., 1970).

Evidence of the presence and impacts of pesticides and other toxic chemicals was encouraging wildlife scientists to pursue research in this direction as well. By the late 1950s traces of toxic chemicals had been identified in numerous species, including fish and fish-eating birds, such as peregrine falcons. Many were alarmed by the massive quantities of DDT sprayed on New Brunswick forests to control the spruce budworm (in 1957 alone 200 aircraft spread over one million kilograms of DDT), as well as by subsequent observations of dead fish and birds (Keenleyside, 1959). The discovery of toxic chemicals such as DDT in the tissues of polar bears and other northern species brought home the ubiquitousness of these invisible threats and helped to generate (after some delay) a political response, including in 1970 a 90 per cent reduction in permitted uses of DDT and within a few years a complete ban.

The CWS quickly assumed leadership in both the science of endangered species, expanding its research to include new species, and the study of chemical threats, particularly pesticides, as well as mercury, oil, and other pollutants. These new directions were closely linked because the impacts of toxic chemicals were felt predominantly not by game species but by other species, such as raptors (including peregrine falcons), that were also potentially endangered. These research areas also, as I explain below, both fit easily into a larger frame of ideas about ecology that was becoming influential.

Not only Canadian scientists, but events elsewhere in the world were encouraging awareness of endangered species. For example, in a few circumpolar nations the polar bear had been thought since the late 1950s to be close to extinction as a result of rapid increases in hunting activity (Stirling and Jonkel, 1972). This concern, particularly as expressed at a 1965 conference in Alaska, focused scientists' attention on the species, encouraging the CWS to maintain a long-term research program to evaluate its status within Canada.

Acting on a larger stage, since 1948 the International Union for Conservation of Nature and Natural Resources (IUCN) had sought to draw the attention of the international community to endangered species (Goodwin, 1973). Canadian scientists, some of them active participants in international affairs, came home from conferences ready to consider such ideas within their own context. Research on pesticides in the United States, Great Britain, and elsewhere also served as models for Canadian researchers, who turned to this question considerably later than did their colleagues elsewhere (Sheail, 1985). Their work intensified after the publication in 1962 of Rachel Carson's *Silent Spring*, which portrayed the impact of pesticides and other toxic chemicals on a wide range of species (Cooch, 1964).

More generally, the influence of ideas from elsewhere was evident in the notion of an ecological ethic based on diversity. Aldo Leopold had written about the role of diversity in maintaining the integrity of ecological systems, while Carson emphasized the impacts of pesticides on all species, not just those of economic significance. Many ecologists argued that the stability of ecosystems was related to their diversity: a simpler ecosystem was also more vulnerable and unstable (Leopold, 1949; Carson, 1962). Scientists in Canada and elsewhere argued that species once thought superfluous in fact had significant roles in ecosystems: Great Lakes fish communities depended on interactions between prey and predator species to maintain their stability, and birds of prey controlled the numbers of other species, helping to maintain an ecological balance (Bocking, 1997; Smith, 1968; Livingston, 1959). Indeed, it could be assumed that all species merited protection because any might play a crucial role in healthy, stable ecological communities. Loss of a species therefore had significance beyond the species itself: it warned of the increasing fragility of ecological systems.

Many scientists and members of the concerned public came to view the loss of species and the impact of toxic chemicals as linked, both serving as signs that the ecological balance of the world might itself be endangered. As the editors of *Canadian Audubon* wrote in a note accompanying one of the first accounts of the widespread impact of DDT in Canada: 'Nature's balance is a delicate and precarious one. We cannot interfere with it lightly. It is quite possible that the indiscriminate spraying of chemical poisons may represent the greatest potential threat to life of all kinds that the world has yet experienced' (note accompanying Keenleyside, 1959: 6).

Leadership of the CWS on endangered species entailed not merely responding to but shaping ideas and public concerns regarding this issue. While some species, such as the trumpeter swan and the whooping crane, had been known to be endangered before the CWS became involved in their study and management, the status of other species only became known through research by the CWS and other agencies. This was the case, for example, with the peregrine falcon, study of which by 1970 had raised concerns that it could become extinct within a decade. In response, in 1972 the CWS established a captive breeding program for raptors at Wainwright, Alberta (Tener, 1973).

As part of its efforts to shape public concerns, the CWS devoted an increasing fraction of its resources to education, through films, including *Atonement*, a 1970 documentary on endangered species, the shorter 'Hinterland Who's Who' items for television (which eventually became a gently satirized element of Canadian pop culture), a wide range of publications, and interpretive work at its wildlife centres, the first of which opened at the Wye Marsh near Georgian Bay in Ontario in 1969 (Tener, 1969; see, generally, Mitman, 1999). Through public education the CWS was able to shape understanding of wildlife while keeping in touch with changing values, helping to ensure that it could respond as these values evolved. Throughout the 1970s the CWS also conducted studies of public perspectives on wildlife, seeking to understand their economic and social significance and how they could be used to demonstrate demand for wildlife protection and for the agency's work (Tener, 1971; Filion and Duwors, 1988).

CWS work on endangered species was in response to the concerns and knowledge of its own scientists, the IUCN and other international agencies, and the public. But it also reflected the institutional landscape of Canadian wildlife agencies during the 1960s. Provincial agencies were developing their own game management and research capabilities. With the formation of territorial wildlife services, the CWS had lost its role as the northern wildlife management agency. Its shift towards non-game species, including endangered species, therefore reflects the development of a new mission for the agency as its older mission was assumed by other agencies. In a sense, the CWS needed endangered species: its reorientation towards new concerns and new species reflected an effort to define fresh areas of activity untended by other agencies. Without these new areas the CWS could not maintain a significant role in Canadian environmental and wildlife affairs. The federal government provided tangible recognition of its success in redefining itself, providing the resources with which the CWS grew rapidly during the 1960s. The provinces also welcomed its research on pesticides and endangered species, supporting its efforts to take an ecological perspective in the face of opposition from powerful interests in agriculture and forestry (Burnett, 1999).

By 1965 the federal government had also begun to revise its own wildlife policies to encompass a wider range of species, habitats, and threats. The 1966 national policy and program for wildlife expressed this broader perspective; among the federal commitments it listed was that of making 'every attempt to prevent any animal from becoming extinct' (CWS, 1967: 5). This policy established the foundations for expanded CWS activities during the following decade, and was followed in 1973 (after several years of discussion) by the Canada Wildlife Act. The Act provided legislative recognition of a broader definition of wildlife, extending beyond game species and migratory birds to include all species and their habitats. It also included a provision permitting the federal government to take steps, in co-operation with the provinces, to protect any non-domestic animal in danger of extinction (Stewart, 1974; Singleton, 1977).

In the early 1970s provincial wildlife departments also, albeit slowly, expanded their attention beyond the bird and game species that had been their traditional focus. A few provinces provided legal recognition for designated endangered species, beginning in 1971 with British Columbia and Ontario. But these initiatives were less than they appeared: as of 1976, BC had yet to designate any species, while the Ontario Endangered Species Act was protecting just four—the peregrine falcon, bald eagle, timber rattlesnake, and blue racer. With some justice, an observer in 1976 described Canadian endangered species legislation and programs as 'piecemeal, jumbled and cosmetic' (Singleton, 1977: 19).

Possibly more significant developments were taking place in the international arena. The 1971 Ramsar Convention for the protection of wetlands, the 1973 Convention on International Trade in Endangered Species (CITES) (which Canada signed in 1975), the 1973 circumpolar agreement on the conservation of polar bears, and restrictions on whale harvesting were among the most prominent obligations regarding endangered species to which Canada assented. Each encouraged further attention to the status of endangered species at home.

New wildlife and endangered species policies were generally seen by government and public alike as part of a larger response to environmental concerns. By 1970 the environment had emerged near the top of the political agenda, as issues as diverse as Great Lakes pollution, the impact of resource development in northern Canada, and preservation of wilderness in British Columbia, Ontario, and elsewhere elicited widespread concern. New public agencies such as Environment Canada and provincial ministries of the environment testified to the emergence of the environment as a public priority.

Citizens also sought organized means through which to express their concerns. Established organizations such as the Federation of Ontario Naturalists expanded, and new groups appeared, such as the National and Provincial Parks Association of Canada (now the Canadian Parks and Wilderness Society [CPAWS]), created in 1963 to encourage the establishment of new parks across the country. International organizations like the World Wildlife Fund spread to Canada, and a multitude of smaller societies expressed people's concerns for their local environment. Advocacy by these organizations drew strength from more general concerns about the environment.

While wildlife science provided essential information regarding the status of endangered species, Canadians' own experience ensured they would prove receptive to this information and to ecological values. Rising prosperity in the postwar era and the availability of more leisure time made the wilderness accessible to a much larger fraction of the population. Outdoor recreation exploded: visits to national parks alone rose from five million in 1960 to 18 million in 1970. As people escaped to parks, arriving on the Trans-Canada and other newly constructed highways, they often overwhelmed the natural and scenic features they had come to enjoy and thus demanded that new parks be created, in part to provide additional opportunities for recreation and wilderness experience. In Ontario, for example, the number of provincial parks grew from eight in 1954 to 108 in 1970.

Historically, Canadian parks had tended to be located where the scenery promised to draw tourists (such as the Rocky Mountains) or in response to local demands or conditions. There was no overall plan to protect ecologically significant features or to ensure that the habitats of endangered species would be protected. By the late 1960s, however, parks designed to include threatened habitats and endangered species began to appear, including Polar Bear Provincial Park in northern Ontario, Gros Morne National Park in Newfoundland, and Kluane National Park in the Yukon.

New rationales for parks reflected at least in part the influence of naturalist and environmental organizations, which advocated a new, ecological basis for choosing parks. This provided another pathway by which changing scientific views of the nation's ecology could influence policy. One of the earliest such advocacy efforts was undertaken by the Federation of Ontario Naturalists. In 1958 it published an 'Outline of a Basis for a Parks Policy for Ontario', which argued that parks be chosen on the basis of protecting significant examples of wildlife and habitat (Killan, 1993).

Parks, the scientists and naturalists argued, should focus not only on maximizing recreational opportunities but on preserving ecological values. A leading figure in efforts to change parks management strategies was Douglas Pimlott (1920-78), professor of zoology at the University of Toronto and a prominent wolf researcher. In his own work he had helped change attitudes towards the wolf by demonstrating both their ecological role and their complex social behaviour (Falls, 1979). Through his activism, he and his colleagues convinced many that parks should be maintained as much as possible in a natural state. In 1968 Pimlott and other concerned individuals formed the Algonquin Wildlands League; over the next several years it helped make logging within parks and the preservation of wilderness significant political issues within Ontario (Killan and Warecki, 1992).

The politics of endangered species shifted radically during the 1960s and early 1970s, encompassing a wider range of species and threats and involving many more individuals, agencies, and issues. By the early 1970s the CWS had been joined by provincial agencies, a wider range of non-government organizations, and a public more interested in wilderness. New ideas about the ecology of species and potential threats furthered this shift. Wildlife management and decision-making were evolving from being conducted within an administrative system closed to most outside influences to being open to participation by a wide variety of interests. Reflecting public interest, 'Endangered Wildlife in Canada' was the theme in 1970 of National Wildlife Week; it was, according to R.C. Passmore of the Canadian Wildlife Federation, 'the most successful of the National Wildlife Week programs to date' (Passmore, 1970: 18).

INSTITUTIONS FOR ENDANGERED SPECIES

In the 1960s endangered species became a political issue; the following decade saw efforts to co-ordinate the diverse agencies and groups acting on this issue. Co-ordination was obviously needed. For example, by the mid-1970s several definitions of 'endangered' and several lists of endangered species were available. (Even Canada Post had its own list, formulated for a series of commemorative stamps.) A single, credible list was necessary to focus attention on species requiring urgent attention and to avoid the controversies often accompanying efforts to identify which were 'really' at risk (Cook and Muir, 1984; Novakowski, 1975).

A more complex conundrum, however, was posed by the intersection of federal and provincial responsibilities for endangered species. Wildlife, like other natural resources, was a provincial responsibility, as was Crown land. Federal responsibility was more circumscribed: migratory birds, national parks, reserves, and the territories. But species and their habitats did not respect provincial borders. Negotiation of international agreements, including CITES, was also a federal matter. The consequence of these tangled jurisdictions was that federal-provincial relations overshadowed almost every discussion or proposal for national action on endangered species. As one observer in the mid-1970s noted regarding an effort by the federal government merely to assemble a list of endangered species: 'it is all very hush-hush because of fear of treading on the toes of the provinces' (Ted Mosquin in Callison, 1977: 182).

While provincial authority put a brake on national initiatives, events outside Canada joined with domestic concerns to compel co-ordinated action. Signed and ratified by Canada, CITES took effect in 1975. It imposed an immediate requirement for Canada to identify endangered species within its borders and to classify them in terms of three categories of varying risk of extinction. There was also now a nearby example of forceful national action: passage by the United States of the 1973 Endangered Species Act led some to ask if a similar initiative would be possible in Canada (Novakowski, 1975).

Well, no, it would not. The obstacles to action in a decentralized federation meant that, even as late as 1975, a whole list of actions had to be taken before, as one official noted, 'we can honestly say we have an endangered species program in, and for, Canada' (ibid., 31). A crucial step forward was provided by a May 1976 symposium that brought together individuals from within and outside government. This symposium gave momentum to an initiative focusing not directly on protecting endangered and threatened species, but on developing a common knowledge basis regarding their status (Mosquin and Suchal, 1977). Its recommendation for a standing committee on endangered species was passed at the Federal-Provincial Wildlife Conference later that year. The following March COSEWIC was formed after a meeting of federal and provincial officials and environmental groups. COSEWIC met for the first time in September 1977 to begin its task of formulating definitions of categories of species at risk and assembling a scientifically credible list of such species.

COSEWIC was a product of its times. Demands from citizens and environmental groups for action (especially evident at the 1976 symposium, which had been sponsored by the Canadian Nature Federation and the World Wildlife Fund Canada) helped force the issue. Provincial resistance to a federal initiative ensured the committee would be composed of officials from every province, as well as from Ottawa and environmental groups, and that it would provide information, not action, with decisions on the status of species carrying no legal consequences. It was also, of course, a response to requirements imposed by CITES. The influence of international examples was also evident in the concept of COSEWIC. The IUCN, unable to act directly in a world of sovereign states, had sought to encourage action on endangered species by providing information, in the form of authoritative reports on the status of species, that could be used in setting conservation priorities (Goodwin, 1973). Acting within a somewhat similar political context of jurisdictions ready to assert their autonomy, this also became COSEWIC's approach: encouraging action by providing information considered credible by all parties.

Operating on very limited resources—'love and a shoestring', in the words of its first chair, J.A. Keith (1979: 74)—COSEWIC had an immediate effect on perceptions and actions regarding endangered species. Within two years its decisions were revealing necessary priorities for research and action. For example, many of the first species identified as being at risk lived only in the short-grass prairie, indicating that this habitat needed urgent protection. (Grasslands National Park in southern Saskatchewan would shortly be formally created.) The committee's appetite for scientific information with which to prepare its status reports also revealed how little Canadian scientists had studied taxa other than birds and mammals, thereby indirectly encouraging research on amphibians, reptiles, and other relatively unknown groups (ibid.). Perhaps most significantly, its reports, tallying a steadily increasing number of species at risk each year, attached urgency to the issue, even if the annual increase reflected more the committee's progress through its systematic survey of the Canadian biota than it did an actual trend in the status of species (Shank, 1999). The COSEWIC list was immediately accepted as authoritative; even Canada Post now adhered to it when choosing endangered species for its stamps (Keith, 1979).

Jurisdictional concerns also influenced federal government activities. This was especially reflected in the work of the Canadian Wildlife Service. In 1978 the CWS created a new endangered species unit to design national programs, co-ordinate regional activities, maintain contact with other agencies, and support the efforts of COSEWIC (Burnett, 1999). This marked a change in emphasis for the CWS: while it would still engage in direct initiatives relating to endangered species, including research and conservation or restoration of bison, peregrine falcons, and a few other species, much of its activity would be devoted to co-ordinating the overall national endangered species effort. With ever more agencies involved, and with government spending restraints now the norm, the CWS could no longer act effectively alone. Even in some fields of wildlife science it was no longer pre-eminent, given the work of university scientists and of organizations like the World Wildlife Fund, particularly in areas such as wildlife toxicology. Accordingly, 'collaboration' and 'co-ordination' would be new watchwords for the CWS.

This orientation would continue into the 1980s, as the CWS began to focus more directly on not merely identifying but restoring endangered species. While the CWS had long engaged in recovery efforts for a few prominent species, by the early 1980s it had begun to plan more strategically its recovery efforts. It proposed a new national strategy to complement COSEWIC by focusing on the recovery of endangered species. RENEW (Recovery of Nationally Endangered Wildlife) was established in 1988 to co-ordinate the recovery efforts of all levels of government and non-governmental organizations (NGOs). Reinforcing its own role as co-ordinator of endangered species activities, the CWS also continued to expand its own efforts, creating in the late 1980s an Endangered Species Division and developing co-operative recovery plans for several species (Burnett, 1999).

It had long been a truism of wildlife management that a strict focus on species, and not their habitats, would prove ineffective. Concerns in the 1950s regarding the viability of northern caribou populations, for example, often focused as much on the supposed impacts of forest fires on their food supply as on losses due to hunting, and until the 1940s reserves formed the centrepiece of northern wildlife conservation. Nevertheless, wildlife research and management, and subsequent action on endangered species, tended to focus on certain species: the whooping crane, bison, peregrine falcon, and a few others. Only in the 1970s did a shift towards habitats and ecosystems become evident. Wildlife scientists and ecologists were becoming increasingly convinced that the well-being of species—and the potential impact of stresses such as forest harvesting, toxic chemicals, and acid rain—could be best understood in terms of ecosystems. Many biologists also became convinced that a strict focus on managing species and populations would be too expensive, and ultimately inadequate (Burnett 1999). Accordingly, CWS biologists as well as scientists within other institutions increasingly shifted efforts towards the study of ecological communities and ecosystems.

Habitat protection was a far more visible political issue than was the relatively quiet process of designation of species at risk. For many Canadians, new parks *were* the issue: the spread of green patches across maps and new opportunities for outdoor experiences and recreation were a far more tangible tally of progress than were endangered species status reports.

During the 1970s and early 1980s initiatives concerning habitats and endangered species developed along separate, yet parallel, paths. Just as COSEWIC substituted a systematic process for the previously haphazard identification of endangered species, so, too, did environmentalists and scientists urge more systematic approaches to identifying protected areas. Between 1968 and 1974, for example, scientists across Canada participated in the Conservation of Terrestrial Ecosystems program of the International Biological Program, identifying and mapping hundreds of areas of ecological significance deserving designation as reserves. In response, a few provinces designated ecological reserves, protecting relatively small areas of outstanding natural and scientific interest. British Columbia led this effort, establishing through its 1971 Ecological Reserves Act 55 reserves totalling 360 square kilometres. Other provinces chose not to establish reserves but used the results of this process in their planning of new parks (McLaren and Peterson, 1975). By 1970 Parks Canada had begun drawing up its own systematic plan for new national parks within each of the natural regions of Canada. Planners used scientific information about ecological and climatic communities to formulate a list of 39 terrestrial and nine marine natural regions that together represented the natural diversity of the nation. The national parks system, it was proposed, would be complete once a representative area within each region had been protected (Carter-Edwards, 1998). And in Ontario environ-mentalists began early in the 1970s to advocate a systematic plan for completing the provincial parks system. In 1978 the province responded with a commitment to com-plete a park system representing the entire natural heritage of the province. Five years later the province announced 155 new parks, adding about two million hectares to the park system (Killan, 1993).

Replacement of the ad hoc designation of endangered species and protected areas by systematic evaluation had its counterpart in the evolution of parks management. By the early 1970s the impacts of increasing numbers of visitors on the nature they had come to enjoy was becoming more evident. In response, managers in several provinces and within the national parks system began to develop strategies for classi-fication and zoning, whereby recreational facilities and activities would be concen-trated within certain parks, or particular areas within parks, leaving other, more eco-logically sensitive parks or areas less exposed. As with the designation of endangered species, a reliance on science in identifying and designating protected areas was expected to result in more effective protection while making decisions more accept-able to affected interests. Wilderness and endangered species became further defined as science-based issues.

Throughout the 1970s progress in protecting habitat was usually measured in terms of the increasing extent of areas owned and protected by governments in parks, eco-logical reserves, and national wildlife areas. By the end of the decade other approach-es to protection were gaining more attention. There were several reasons for this. Budget cuts were affecting government's ability to acquire land; in particular, by the late 1970s the CWS acquisition program had been severely reduced. Controversies over land purchases and displacement of communities to make way for national parks (notably in New Brunswick and Newfoundland) created a backlash against land acquisitions. Resource conflicts added more complications. For example, the new Grasslands National Park in southern Saskatchewan took far longer to be estab-lished (in 1981) than had earlier national parks in western Canada. Surveys and plan-ning exercises had also identified the need for protected areas in more highly popu-lated regions, such as southern Ontario, where within the Great Lakes/St Lawrence Life Zone many rare species existed on the northernmost fringe of their distribution. Land there was also mostly privately held and often very expensive. As a result of all these factors, alternatives to government acquisition received more attention. Foremost was the notion of stewardship, in which landowners would be encouraged to manage their land as habitat for wildlife, with government providing advice and, occasionally, legal protection and economic incentives.

This emphasis on co-operative habitat protection was reinforced by the creation in 1984 of Wildlife Habitat Canada, which had a mandate to conserve, restore, and enhance wildlife habitat. A joint federal, provincial, and territorial agency with repre-sentation from NGOs and the private sector, it promoted stewardship (Burnett, 1999). A similar orientation was evident in the North American Waterfowl Management Plan, a massive project begun in 1986 by the American and Canadian governments

(Mexico also signed on in 1994) to increase waterfowl habitat. Also participating were provincial and state governments, NGOs (led by Ducks Unlimited), and private landowners. While focusing primarily on waterfowl of interest to hunters, the plan has more recently attempted to broaden its focus to include other species (Baydack, 2000).

SPACES, SPECIES, AND BIODIVERSITY

By the late 1980s interest in species and their habitats had begun to evolve towards a more broadly defined concern with all life on earth. The concept of 'biodiversity' came to the attention of Canadians primarily as a result of its emergence as a unifying theme in international environmental affairs, combining elements of both science and politics: an appreciation of the vast diversity of life and of the combination of initiatives necessary to safeguard it. The term 'biodiversity' itself achieved prominence at a 1986 conference in Washington, DC (Wilson, 1988). A Convention on Biological Diversity was a major product of the 1992 United Nations Conference on Environment and Development. In 1995 the federal government released its Canadian Biodiversity Strategy as its official response to the biodiversity convention.

The 1992 convention exemplified the extent to which conservation issues, including endangered species, had been subsumed within the broad issue of biodiversity. In Canada as well, wildlife policy has assumed a broader stance, encompassing concerns regarding habitat, endangered species, and, indeed, all species. In 1990 a new national wildlife policy, encapsulating ideas long in circulation, expressed a federal commitment to protect species and their habitats (Wildlife Ministers Council of Canada, 1990). The following year a National Wildlife Strategy was introduced, which included the development and implementation of recovery plans for species at risk, as well as increased support for co-operative management of habitat and the expansion of volunteer-based non-game bird studies (Burnett, 1999). These initiatives exemplified how ideas about wildlife had changed, both in terms of a more inclusive definition of species of interest and in terms of an acknowledged need for broad participation in their protection. At the same time, restrictive definitions of some programs (RENEW, for example, has until recently been limited to terrestrial vertebrates) illustrate the unevenness of this evolution towards more inclusive definitions.

Of particular note in the politics of endangered species of the 1990s was the extent to which leadership had been assumed by NGOs. At the local level many small organizations act as stewards and protectors of valued habitats (DeMarco and Bell, 2000). Nationally, in September 1989 World Wildlife Fund Canada and the Canadian Parks and Wilderness Society launched the Endangered Spaces campaign, with the goal of protecting at least 12 per cent of Canada's land area by the year 2000. This goal was inspired by the World Commission on Environment and Development, which recommended in 1987 that all nations commit to protecting 12 per cent of their land area (WCED, 1987). While the 12 per cent figure was a convenient symbolic goal, the Endangered Spaces campaign soon redefined its objectives in more ecologically defensible terms, emphasizing the need for protected areas within each of the ecological regions of Canada. The following year Lucien Bouchard, at the time the federal Environment Minister, committed his government to completing the national park system, including five new national parks on land by 1996 and agreements on 13 more by 2000. Each year since then the World Wildlife Fund has provided annual report cards marking progress made by federal and provincial governments towards these targets. Progress has been slow, and the year 2000 target was not reached.

In parallel with the Endangered Spaces campaign, environmental groups also focused national attention on the need for stronger legislative protection for endangered species, particularly as a way of implementing the Biodiversity Convention. In 1992 the World Wildlife Fund, Canadian Nature Federation, Canadian Parks and Wilderness Society, Sierra Club, and Canadian Environmental Law Association submitted a brief to the Parliamentary Standing Committee on the Environment, urging endangered species legislation. This led to a 1995 federal paper, 'The Canadian Endangered Species Protection Act: A Legislative Proposal', a National Accord for the Protection of Wildlife Species at Risk in Canada (approved in 1996), and the drafting of a Canadian Endangered Species Protection Act. Lengthy hearings, controversy, the demise of the proposed legislation, its reappearance (with a more inclusive concern for species at risk) in April 2000, and its demise, again as a result of an election call, have marked its most recent history.

CONCLUSION

The history of endangered species politics can be traced in a sequence of events: creation of national parks, identification of species and habitats at risk, establishment of COSEWIC and other institutions, negotiation of international agreements. Often, however, the absence of events has been equally significant. Soon after it took place, the 1976 Symposium on Canada's Endangered Species and Habitats was cited as helping to form 'a sound basis for a truly national endangered species program' (Loughrey, 1977: 47). Nearly a quarter-century later, Canada is only now, perhaps, about to gain such a program. One is struck by the episodic nature of this history: the rapid evolution of the political context of endangered species between the late 1960s and early 1970s, followed by 25 years of tentative pursuit of possibilities defined during that brief period of change.

An understanding of these possibilities, and their uncertain expression, can be achieved by considering broader trends in the history of the politics of endangered species. These include the expansion of concerns regarding species, from a few prominent animals and birds to a more inclusive perspective encompassing most species (although certain charismatic species continue to attract most attention). This was paralleled by awareness of a broader range of impacts on wildlife: from hunting to habitat loss, toxic chemicals, and other pollutants. There was also a broader range of actors involved in endangered species. In the 1950s the CWS dominated; by the 1990s wildlife was of interest to many more—provincial governments, environmental organizations, landowners, and the general public. This broadened base of interest generated a greater need for partnerships and for collaborative entities such as

COSEWIC. International developments, including CITES and other treaties, and the Convention on Biological Diversity, have also influenced Canadian endangered species politics.

The emergence itself of endangered species politics presents an interesting historical problem, particularly since many of these species have little or no economic value. Why have these species elicited so much concern?

Part of the answer lies in the influence of scientists and naturalists. Research has played a major role, providing an expanding knowledge base regarding species and their ecological relations, identifying the need of many species for large areas of habitat, recognizing the significance of corridors, fragmentation, and other characteristics of habitats, and providing knowledge concerning the status of specific populations and the significance of human impacts, both traditional (hunting) and more novel (such as toxic chemicals and climate change). The role of science in providing a basis for managerial and political decisions, such as regarding the status of species and the designation of new parks, has also been significant. And environmental activists have often urged a stronger role for science as a way of enhancing the defensibility of plans for protection and to limit government discretion. The influence of prominent scientist/activists, such as Douglas Pimlott, has been noteworthy. To some extent, then, the emergence of endangered species can be interpreted in terms of the ideas of scientists and the dissemination of these ideas into society, where they have served both as inspiration and as a political resource.

However, this history cannot be understood merely in terms of the initiatives of an élite group of scientists and other opinion leaders. Their influence was often limited, and several arenas in which they were most active, such as COSEWIC, have remained almost unknown to most Canadians. Instead, these events and trends must be placed within the context of the evolving relationship between Canadians and their environment. This relationship has been marked by the increasing impact of human activities. Resource industries—forestry, fishing, mining, as well as agriculture—have reshaped large areas, eliminating habitat for many species (Mosquin, 2000). Exotic species have displaced native species. The movement of large numbers of people into the countryside, to build cottages and recreational facilities, has reversed an earlier trend of rural depopulation and has resulted in fragmented habitats, producing conditions favourable to species able to coexist with humans and forcing out those requiring larger undisturbed areas. More subtle and more pervasive impacts—toxic chemicals, acid rain, depletion of the ozone layer, climate change—have threatened to degrade habitats or decimate populations. These impacts have together encouraged widespread public concern regarding the loss of species and habitats.

But the emergence of widely held concerns regarding wilderness and endangered species can also be understood in terms of broader changes in social values, reflecting not only the dissemination of ideas from an élite and direct experience of environmental damage, but also changes in the experience and values of society as a whole, particularly in terms of how people use, understand, and think about the natural environment. These changes involve a search for a better quality of life. Economic and social changes in society over the last 50 years, including the declining economic significance of resource production, rising standards of living, and increased leisure time, have encouraged people to develop a desire for amenities, including opportunities to experience wilderness and wildlife, together with other amenities, such as clean air and water, that are only available through a protected environment. Endangered species politics, therefore, reflects widespread acceptance of the idea that wildlife, and the environment as a whole, can have non-economic significance to humans, whether this significance is expressed in terms of aesthetic, ecological, intrinsic, or other values.

Thus, while scientists and other opinion leaders provided new information about other species and our impacts on them, changing public values, rooted in people's own lives and experience, provided much of the political impetus behind the creation of new parks and other protected areas and for the protection of wildlife. Often, people found rural areas and wilderness to be congenial places in which to express these new environmental values, as was reflected in more visits to national parks, the proliferation of cottages, and a surge in outdoor recreation. These values were also expressed through a growing interest in land stewardship. In contrast, more traditional ways of experiencing wilderness, through consumption in the form of hunting and trapping, have continued to decline in significance.

As with any long-term change in societal values, newer environmental values have existed alongside older perspectives that view the landscape in terms of resources to be consumed. Conflict between these new and old values has, as a result, been a central element in controversies over endangered species, delaying actions and initiatives once seen as inevitable. The history of endangered species politics suggests that such conflicts will continue. While economic activity and attitudes based on resource extraction remain deeply rooted in Canada, so, too, now are the new environmental values that have compelled change in our relations with other species.

REFERENCES

Banfield, A.W.F. 1956. 'The Caribou Crisis', *Beaver* 286: 3–7.

Baydack, R.K. 2000. 'Science and Biodiversity', in Bocking (2000: 175–87).

Bocking, S. 1997. 'Fishing the Inland Seas: Great Lakes Research, Fisheries Management and Environmental Policy in Ontario', *Environmental History* 2: 52–73.

——, ed. 2000. *Biodiversity in Canada: Ecology, Ideas, and Action.* Peterborough, Ont.: Broadview Press.

Burnett, J.A. 1999. 'A Passion for Wildlife: A History of the Canadian Wildlife Service, 1947–1997', *Canadian Field-Naturalist* 113: 1–183.

Callison, C.H. 1977. 'Endangered Species and Habitats: The Role of Non-Governmental Organizations', in Mosquin and Suchal (1977: 182–3).

Canadian Wildlife Service. 1967. *Canadian Wildlife Service '66.* Ottawa.

Carson, R. 1962. *Silent Spring.* Boston: Houghton Mifflin.

Carter-Edwards, D. 1998. 'The History of National Parks in Ontario', in J. Marsh and B. Hodgins, eds, *Changing Parks: The History, Future and Cultural Context of Parks and Heritage Landscapes.* Toronto: Natural Heritage/Natural History, 94–106.

Cooch, F.G. 1964. 'Current Developments in the Biocide-Wildlife Field', *Canadian Audubon* 26: 148–50.

Cook, F.R., and D. Muir. 1984. 'The Committee on the Status of Endangered Wildlife in Canada (COSEWIC): History and Progress', *Canadian Field-Naturalist* 98: 63–70.

DeMarco, J.V., and A.C. Bell. 2000. 'The Role of Non-Government Organizations in Biodiversity Conservation', in Bocking (2000: 347–65).

Department of Northern Affairs and National Resources. 1956. *Minutes of the 20th Federal-Provincial Wildlife Conference*. Ottawa.

Erskine, A.J. 2000. *Birds in Boreal Canada: Communities, Densities and Adaptations*. Ottawa: Canadian Wildlife Service.

Falls, J.B. 1979. 'Douglas H. Pimlott: Lessons for Action', *Nature Canada* 8, 2: 18–23.

Filion, F.L., and E. Duwors. 1988. 'Measuring the Demand for Non-Hunting Uses of Wildlife', *Transactions of the 52nd Federal-Provincial/Territorial Wildlife Conference*: 66–83.

Goodwin, H.A. 1973. 'Ecology and Endangered Species', *Proceedings of the 38th North American Wildlife Conference*: 46–55.

Harington, C.H. 1964. 'Polar Bears and Their Present Status', *Canadian Audubon* 26: 4–11.

Keenleyside, M.H.A. 1959. 'Insecticides and Wildlife', *Canadian Audubon* 21: 1–7.

Keith, J.A. 1979. 'Committee on the Status of Endangered Wildlife in Canada', *Transactions of the 43rd Federal-Provincial Wildlife Conference*: 71–4.

Killan, G. 1993. *Protected Places: A History of Ontario's Provincial Parks System*. Toronto: Dundurn Press.

—— and G. Warecki. 1992. 'The Algonquin Wildlands League and the Emergence of Environmental Politics in Ontario, 1965–1974', *Environmental History Review* 16, 4: 1–27.

Leopold, A. 1949. *A Sand County Almanac*. Oxford: Oxford University Press.

Lewis, H.F. 1947. 'Wildlife Conditions in Canada', *Transactions of the 12th North American Wildlife Conference*: 10–16.

Livingston, J.A. 1959. 'Can We Save the Birds of Prey', *Canadian Audubon* 21: 58–9.

Loughrey, A.G. 1977. 'Canadian Wildlife Service', *Transactions of the 42nd Federal-Provincial Wildlife Conference*: 40–52.

McAllister, D.E., F.R. Cook, N.S. Novakowski, and W.E. Godfrey. 1970. 'Rare or Endangered Canadian Fishes', 'Rare or Endangered Canadian Amphibians and Reptiles', 'Endangered Canadian Mammals', and 'Canada's Endangered Birds', *Canadian Field-Naturalist* 84, 1: 5–26.

McLaren, I.A., and E.B. Peterson. 1975. 'Ecological Reserves in Canada: The Work of IBP-CT', *Nature Canada* 4, 2: 22–31.

Mitman, G. 1999. *Reel Nature: America's Romance with Wildlife on Film*. Cambridge, Mass.: Harvard University Press.

Mosquin, T. 2000. 'Status of and Trends in Canadian Biodiversity', in Bocking (2000: 59–79).

—— and C. Suchal, eds. 1977. *Canada's Threatened Species and Habitats: Proceedings of the Symposium on Canada's Threatened Species and Habitats, May 20–24, 1976*. Ottawa: Canadian Nature Federation.

Munro, D.A. 1962. 'Trumpeter Swans', *Canadian Audubon* 24, 3: 65–9.

North American Wildlife Conference. 1942. 'Report on Game Conditions in the Canadian Provinces', *Transactions of the 7th North American Wildlife Conference*: 12–17.

Novakowski, N.S. 1975. 'Endangered Species', *Transactions of the 39th Federal-Provincial Wildlife Conference*: 28–31.

Page, B. 2000. 'Banff National Park: The Historic Legacy for Biodiversity', in Bocking (2000: 31–56).

Passmore, R.C. 1970. 'Report of the Canadian Wildlife Federation', *Transactions of the 34th Federal-Provincial Wildlife Conference*: 18–20.

Pimlott, D.H. 1961. 'Wolf Control in Canada', *Canadian Audubon* 23, 5: 145–52.

Shank, C.C. 1999. 'The Committee on the Status of Endangered Wildlife in Canada (COSEWIC): A 21-Year Retrospective', *Canadian Field-Naturalist* 113: 318–41.

Sheail, J. 1985. *Pesticides and Nature Conservation: The British Experience 1950–1975*. Oxford: Clarendon Press.

Singleton, M. 1977. 'Endangered Species Legislation in Canada', in Mosquin and Suchal (1977: 19–21).

Smith, S.H. 1968. 'Species Succession and Fishery Exploitation in the Great Lakes', *Journal of the Fisheries Research Board of Canada* 25: 667–93.

Stewart, D. 1974. *Canadian Endangered Species*. Toronto: Gage.

Stirling, I., and C. Jonkel. 1972. 'The Great White Bears', *Nature Canada* 1: 15–18.

Tener, J.S. 1960. 'The Present Status of the Barren-ground Caribou', *Canadian Geographical Journal* 60: 98–105.

——. 1969. 'Report of the Director of the Canadian Wildlife Service', *Transactions of the 33rd Federal-Provincial Wildlife Conference*: 17–18.

——. 1971. 'Report of the Canadian Wildlife Service', *Transactions of the 35th Federal-Provincial Wildlife Conference*: 13–16.

——. 1973. 'Report of the Canadian Wildlife Service, 1973', *Transactions of the 37th Federal-Provincial Wildlife Conference*: 12–18.

Wildlife Ministers Council of Canada. 1990. *A Wildlife Policy for Canada*. Ottawa: Canadian Wildlife Service.

Wilson, E.O., ed. 1988. *Biodiversity*. Washington: National Academy Press.

World Commission on Environment and Development. 1987. *Our Common Future*. Oxford: Oxford University Press.

In Search of a Minimum Winning Coalition: The Politics of Species-at-Risk Legislation in Canada

WILLIAM AMOS, KATHRYN HARRISON, AND GEORGE HOBERG

In 1992, Canada was the first industrialized country to sign the Convention on Biological Diversity in Rio de Janeiro. Yet as we write in November 2000, there is still no Canadian federal statute designed to protect species at risk.[*] Although some authority to protect species at risk can be found in existing federal statutes, and provincial governments also provide varying levels of protection to species at risk, the current legislative framework is an inconsistent and incomplete patchwork of federal and provincial laws. The absence to date of national species-at-risk legislation is particularly striking if one considers that the United States adopted its Endangered Species Act almost three decades ago, in 1973.

The lack of federal legislation is not entirely for want of trying, at least in recent years. The Liberal government proposed its Bill C-65, the Canada Endangered Species Protection Act, in late 1996, only to have it die on the order paper amid opposition from all sides when a federal election was called in the summer of 1997. The Liberal government subsequently introduced the Species at Risk Act (Bill C-33) in April 2000. But again the bill died on the order paper when the Chrétien government called the 2000 election. The chapter seeks to explain this delay in adoption of national endangered species legislation in Canada and, in particular, what the failure of Bill C-65 can tell us about the politics of species-at-risk protection in Canada and the prospects for success of future legislation. The analysis focuses on the influence of ideas in the form of scientific knowledge, organized interests, and institutions. Although we do not offer a full-blown comparative study, we draw occasional comparisons to the US experience to inform our understanding of species-at-risk policy-making in Canada.

In short, Bill C-65 did not succeed because the federal government was unable to construct a coalition of political support large enough to permit passage. For environmentalists, the bill did not go far enough in protecting the habitat of species at risk. For industry and landowners, the bill went too far in the direction of an aggressive and potentially costly regulatory approach. For provinces, the bill threatened too

---

[*] Any views or comments, actual or inferred, contained in this article are solely those of the authors and should not be attributed to any other individual or organization.

much federal intervention. Finally, First Nations objected that they were not adequately consulted and that their authority to manage their own resources was not adequately recognized. With vehement criticism from all sides, the bill died. Although the federal government has clearly learned from that experience and attempted to craft compromise legislation that can win over at least some players from each camp, it remains to be seen whether another attempt can surmount the considerable institutional and political obstacles to federal endangered species legislation.

THE POLICY REGIME FRAMEWORK

In analysing the failure of Bill C-65, this chapter employs a policy regime framework that seeks to explain policy outcomes as a function of three regime components: actors, institutions, and ideas (Cashore et al., forthcoming; Hoberg and Morawski, 1997). The regime approach considers not only the independent causal effects of each factor on policy outcomes, but also their interactions in the context of background conditions. The first regime component, *actors*, is defined as the individuals and organizations, both public and private, that play an important role in the formulation and implementation of public policies. The pursuit of interests within a competitive political arena is structured both by the resources that each actor can draw upon to influence policy outcomes and by the strategies employed to maximize the impact of those resources.

The institutional conditions under which this competition of interests occurs represent the second element of the regime. *Institutions* structure the roles, interactions, and occasionally the interests of legitimate policy participants, both governmental and non-governmental. Institutions can influence the policy-making process by allocating authority and shaping relations between government actors (i.e., between the executive, the legislature, and the judiciary, as well as between federal and provincial governments), and by defining the rules of and opportunities for participation by individuals and groups.

The third regime component, *ideas*, is defined here to include both causal and normative beliefs about the substance and process of public policy. Ideas can serve as political resources for non-governmental actors, and thus present both constraints and opportunities for government decision-makers. Policy-makers may also be independently motivated by their own normative ideas or causal knowledge about policy problems and solutions.

The independent and interactive influence of these three regime components—actors, institutions, and ideas—can only be understood in a specific context. Contextual factors can include economic conditions, simultaneous developments in other policy arenas, international relations, and public opinion (to name but a few). Changes in background conditions frequently change the allocation of resources among regime actors, shifting the balance of power.

Within the context of institutions, ideas, and broader background conditions, actors pursue their interests by adopting the strategies they believe are most likely to succeed in advancing those interests. Government actors, those with the authority to make public policy, confront the challenge of constructing sufficient support within their political environment to enable policies to be enacted and implemented. In the complex circumstances of Canadian environmental policy, there are no hard and fast rules as to what constitutes a 'minimum willing coalition'. As we will see, the federal government's first major effort to enact comprehensive species-at-risk legislation failed that test miserably. We will return to the question of what constitutes a winning coalition in our discussion of C-33 in the epilogue.

SPECIES-AT-RISK LEGISLATION IN CANADA

The Legislative Context

The listing of species at risk in Canada has been undertaken since 1978 by the Committee on the Status of Endangered Wildlife in Canada (COSEWIC). Composed of scientific experts from provincial, territorial, and federal wildlife agencies, as well as from non-governmental conservation organizations, COSEWIC is the recognized Canadian authority on species assessment. However, as federal and provincial legislation currently stands, the COSEWIC list has no legal authority and no regulatory consequences.

A patchwork of federal statutes provides some authority for protecting species at risk. Both the Fisheries Act and the Migratory Bird Convention Act authorize measures to protect some species and their habitats. However, neither statute contains provisions specifically concerning species at risk, nor do they call for development of strategies to protect such species. The National Parks Act authorizes imposition of penalties for individuals who 'hunt, disturb, confine, or possess threatened or protected species', but only in the context of national parks. Similarly, the Wild Animal and Plant Protection and Regulation of International and Interprovincial Trade Act, enacted in 1992 in response to the international Convention on International Trade in Endangered Species (CITES), prohibits the international or interprovincial transport of endangered or threatened plants and animals (and their products), but does not extend to aspects of species-at-risk protection other than trade and commerce.

Finally, the 1973 Canada Wildlife Act authorizes the federal government to undertake public education and research on wildlife conservation and to work in co-operation with the provinces to take measures for the protection of endangered wildlife species. The Act provides broader mechanisms for protecting endangered plant and animal species than other statutes, including the acquisition and management of wildlife habitat. However, protection of species at risk is not the central focus of the Act. It thus does not establish a legal foundation for the COSEWIC (or any other) list of species at risk, direct the minister to develop federal-provincial strategies, clarify conditions under which unilateral federal plans can be pursued, or provide guidance as to the nature and scope of such plans.

In regard to provincially controlled lands, the protection for species at risk varies from coast to coast. As owners of Crown lands, the provinces have clear and extensive proprietary authority over flora and fauna, including species at risk, on those lands.

Since 1996, six provinces and territories have introduced new or improved species-at-risk legislation, in addition to the four provinces that had already enacted legislation. However, provincial laws vary in terms of the stringency of prohibitions, extent of habitat protection, and degree of enforcement. The federal pattern of patchwork statutes is repeated in other provinces. For instance, like the federal government, British Columbia currently relies on an 'umbrella' of statutes and programs, including the Forest Practices Code, the Wildlife Act, and the Protected Areas Strategy. As in the federal case, development and implementation of recovery plans are not mandatory, and the level of protection authorized is inconsistent across species.

In the absence of a federal endangered species law, and given the diversity of provincial statutes and policies, varying levels of protection are currently provided for Canada's species at risk. It is particularly noteworthy that existing federal and provincial statutes are all discretionary; they authorize some actions but do not mandate any particular actions to protect species at risk. The piecemeal and discretionary nature of Canada's legislative framework is especially apparent when contrasted with the federal Endangered Species Act (ESA) in the United States. The ESA requires the designation of both 'endangered' and 'threatened' species based on the best available scientific information. Upon listing, species and their critical habitats receive strong protection in the form of broad prohibitions of actions that might harm the species, including significant modification of habitat, as well as mandatory development and implementation of recovery plans. The statute's prohibition on 'taking' a species or its habitat applies to both government and private actors. Moreover, citizen suits can be brought by any person against anyone alleged to be in violation of the Act. Combined with the non-discretionary language of the statute (the ESA does not simply *authorize* protection, it *compels* it), these civil suit provisions have led to litigious and often controversial enforcement of the ESA.

The History of Bill C-65

The United Nations Conference on the Environment and Development, held in June 1992 in Rio de Janeiro, marked the beginning of efforts to develop Canadian federal species-at-risk legislation. Amid great political fanfare, Canadian representatives returned from the Rio Summit as the first industrialized country to sign the Convention on Biological Diversity. Under Article 8k of the Convention, signatory countries are obliged to 'develop or maintain necessary legislation and/or other regulatory provisions for the protection of threatened species and populations' (United Nations, 1992: 8). When the Mulroney government ratified the convention in December 1992, it committed itself to enacting stand-alone species-at-risk legislation.

In early 1993, the House of Commons Standing Committee on the Environment held hearings, and subsequently recommended that Environment Canada initiate a comprehensive intergovernmental approach to protect species at risk and their habitats. Following the Standing Committee's report, a multi-stakeholder focus group co-ordinated by Canadian Wildlife Service officials began the task of figuring out how an integrated national approach to wildlife management might work. Their report recommended that both provincial and federal governments should enact 'comprehensive legislation . . . to ensure the protection of species, ecological communities, and ecosystems', and that the 'federal government should frame national minimum standards for designation and protection of endangered species . . . and their habitats' (Canadian Wildlife Service, 1993: 2–3).

With Deputy Prime Minister Sheila Copps as Environment Minister in the newly elected Liberal government, a multi-stakeholder Task Force on Endangered Species Conservation was established in April 1994. This Task Force of representatives from major resource industries, private landowners, academia, and environmental groups provided extensive policy recommendations to the federal government in its final report in May 1996. The federal government also pursued other concurrent avenues for consultation. In November 1994 the government circulated a discussion paper among stakeholders and invited public comments on the legislative proposals contained therein (Environment Canada, 1994). Regional consultation workshops were held across Canada in May 1995, through which both the federal and provincial governments co-ordinated their efforts to gain stakeholder input concerning the discussion paper. Thereafter, in the fall of 1995, the minister released a proposal for a species-at-risk bill (Environment Canada, 1995).

On 2 October 1996 Canada's federal and provincial wildlife ministers reached an agreement in the Wildlife Ministers Council of Canada, which committed each jurisdiction to pass legislation and programs protecting species at risk and their habitats. Developed in the spirit of an ongoing federal-provincial environmental 'harmonization' exercise (Harrison, 2000a), the National Accord for the Protection of Species at Risk was an intergovernmental framework designed to ensure a seamless web of protective legislation across all Canadian jurisdictions. Following this announcement, the new federal Environment Minister, Sergio Marchi, introduced Bill C-65, the Canada Endangered Species Protection Act (CESPA), in the House of Commons on 31 October 1996.

The main provisions of Bill C-65 (as well as provisions of the subsequent Bill C-33 and the positions of key interest groups) are summarized in Table 7.1. While COSEWIC would continue its role in analysing species at risk, the decision about listing would be made by the Minister of Environment. The scope of application of the bill was relatively narrow: it covered plant and animals species on federal lands (but not provincial or private lands), aquatic species, and migratory birds covered by the Migratory Birds Convention Act. Once listed as endangered or threatened, the species could not be killed, nor could the 'residence' of the species, defined as being a nest or den, be destroyed. Recovery plans, to be developed in consultation with stakeholders, would be required within one year for endangered species and within two years for threatened species. The bill would authorize cost-sharing arrangements. Finally, the bill proposed a novel citizen suit provision, authorizing citizens to sue the government for failing to perform non-discretionary duties under the law. Marchi proclaimed that the 'new Act—plus the Accord—equals protection for endangered

Table 7.1: Summary of Positions on Canadian Species-at-Risk Legislation

	C-65	Key Committee Changes to C-65	Environmental NGOs	Industry and Landowners	Provinces	C-33
Listing	COSEWIC recommends listing, cabinet decides.	No change.	Want scientific committee, not politicians, to decide what gets listed.	Favour cabinet discretion.	Wanted provincial consent to COSEWIC appointments.	COSEWIC recommends listing, cabinet decides. Greater provincial role in COSEWIC appointments.
Scope of application	Requires protection of listed species on federal lands, aquatic species and migratory birds everywhere. Authorizes discretionary protection of international trans-boundary species and their residences everywhere.	Made federal protection of listed international trans-boundary animals and their residences mandatory, except when covered by equivalency agreements with provinces.	Want federal law to cover all listed species everywhere in Canada.	Oppose federal intervention on international trans-boundary species on provincial lands.	Oppose federal intervention on international trans-boundary species on provincial lands.	Broadens scope to include individuals, residences, and critical habitat of all listed species anywhere in Canada.
Protection	Bans killing listed species or destroying their 'residences', defined as 'nest or den'. Requires development of recovery plan within one year for 'endangered' and within two years for 'threatened' species.	Expanded definition of 'residences' to include breeding, rearing, and hibernating areas.	Want protection of critical habitat in addition to individuals and their residences.	Generally support less expansive definitions.	Oppose federal protection of habitat inasmuch as that entails federal 'intrusion' on provincial lands.	Authorizes protection of individuals, residences (more broadly defined than in C-65), and 'critical habitat'. Introduces two-stage process to develop recovery plans and action plans. Similar deadlines for recovery plans as in C-65.

Table 7.1 — continued

	C-65	Key Committee Changes to C-65	Environmental NGOs	Industry and Landowners	Provinces	C-33
Federal-provincial arrangements	Defers to provincial governments other than on federal lands and with respect to migratory birds and aquatic species. Defers to provinces with respect to international species at first instance.	Greater emphasis on federal role in making protection of international trans-boundary species mandatory except where covered by equivalency agreements.	Distrust federal deference to provinces, arguing it will result in 'patchwork' protection.	Support federal-provincial co-operation, 'one window' implementation.	Argue that provinces should take lead role; opposed to equivalency approach.	Authorizes federal involvement to protect individuals, residences, and critical habitat on provincial and territorial lands only if the subnational government is not doing the job. No equivalency provisions.
Discretion and citizen suits	Authorizes citizens to sue federal government to force non-discretionary actions.	Reduced federal discretion with respect to trans-boundary species.	Favour automatic prohibitions across all jurisdictions, and citizen suits.	Oppose adversarial and litigatory regulatory approach in favour of more co-operative approach; oppose citizen suits in light of US experience.	Oppose citizen suits; oppose non-discretionary duties that require federal 'intrusion' in provincial jurisdiction.	No citizen suit provisions. Discretion is, if anything, greater than in Bill C-65.
Costs	Authorizes expenditures and cost-sharing, but does not emphasize funding of stewardship programs or mention compensation for regulatory takings.	No change.	Support compensation.	Want guarantees of compensation.	Nervous about implications of federal compensation in setting a precedent for provincial programs.	Emphasizes funding for stewardship programs; authorizes compensation for regulatory takings.

species from coast to coast to coast. We now have a comprehensive national frame-work for action—not a patchwork quilt' (Office of the Minister of the Environment, 1996: 1).

Despite Marchi's enthusiasm, Bill C-65 was attacked by virtually every stakeholder group—industry, landowners, environmentalists, First Nations, and provincial governments. Business and landowners felt that the federal government had gone 'too far' with this proposed legislation, giving endangered species priority over people. They were joined by provincial governments and First Nations, who questioned the federal government's authority to pursue its proposals. At the same time, the environmental and scientific communities claimed the proposal did not go nearly far enough. They labelled the proposed legislation constitutionally timid and inadequate to provide meaningful protection for species at risk. Clearly, where one stood on Bill C-65 depended on where one sat: it was either too strong or too weak, too punitive or riddled with too many loopholes, jurisdictionally overbearing or too limited in scope. Stakeholders from all sides criticized the federal government for ignoring their recommendations from the consultation process.

After first reading, the notoriously independent Standing Committee on the Environment and Sustainable Development, chaired by Charles Caccia, held several weeks of public hearings and reported a stronger version of the bill back to Parliament on 3 March 1997. (Key committee amendments are also summarized in Table 7.1.) Cabinet then spent several weeks developing a series of additional proposed amendments to the bill. The 115 proposed amendments tabled in the House of Commons (40 of which were proposed by the Environment Minister) would have diluted some of the bill's environmental strengths, many of which had been introduced by Caccia's committee. Although debate on this package of amendments began in the House of Commons on 24 April 1997, Bill C-65 was placed on the legislative backburner amid rumours of a spring election. When Parliament was dissolved prior to a June election, the final version of Bill C-65 died on the order paper.

PUBLIC OPINION AS A CONTEXTUAL FACTOR

Although background conditions shaping the political environment include such factors as economic conditions and concurrent developments in other policy fields, by far the most important background condition in this case was public opinion. Public opinion was crucial because it determined which issues re-election-minded politicians paid attention to and which direction they were likely to go. In Canada, public opinion on species at risk has been characterized by very high support for protective legislation, but with relatively low priority on the issue—a classic case of support being a kilometre wide but a centimetre deep.

Various authors have emphasized the importance for environmental policy of the cyclical nature of public attention to the environment (Harrison, 1996; Hoberg, 1993; Doern and Conway, 1995). Within Canada, the most significant environmental policy initiatives at both the federal and provincial levels have occurred during the peaks of public concern for environmental issues. Thus, with regard to federal species-at-risk legislation, one might expect that where we are in the public opinion cycle could significantly influence the balance of political forces and the likelihood of federal legislation (as well as its substance). The fact that Bill C-65 was proposed during a 'trough' in public opinion strengthened the position of opponents of species-at-risk legislation and undermined that of advocates.

On one level, it is not surprising that there is strong support among Canadian voters for legislation to protect species at risk. Protecting the environment, including endangered species, is a 'motherhood' issue that voters invariably support. Environmentalists have repeatedly emphasized the high degree of electoral support for federal species-at-risk legislation revealed by public opinion polls. In particular, a May 1995 Angus Reid poll found that 94 per cent of those polled responded positively when asked the following question:

> According to experts, there are 243 endangered, threatened and vulnerable species of animals and plants in Canada, such as the Beluga Whale and the Prairie Rose. Presently, these species have no national legal protection, although the federal government is currently considering such legislation. Would you support or oppose federal legislation that would protect endangered species? (Angus Reid Group, 1995)

A subsequent Pollara poll, commissioned by the International Fund for Animal Welfare (IFAW) in May 1999, found that 92 per cent supported federally established 'national standards [for endangered species protection] that would apply in all provinces and territories' (IFAW, 1999). These polls paint a picture of a Canadian electorate that is nearly unanimous in its support for strong federal legislation to protect species at risk.

There are, however, a number of reasons to question the depth of this support. First, it is important to recognize the ignorance of the average voter on this issue. Not surprisingly, IFAW chose not to publicize the fact that 66 per cent of the Pollara respondents thought the federal government already had a law to protect endangered species, and another 21 per cent did not know whether such a law existed (Pollara, 1999: 30). A second issue relates to the design of these surveys, which alerted respondents to a problem—exemplified by the Angus Reid preamble emphasizing both the magnitude of threats to species at risk and the lack of governmental response—but not to the sacrifices that might be called for to address it. With no understanding of the complexities of this legislative issue, who could say no to an imperiled beluga or prairie rose?

Finally, however genuine electoral support for species-at-risk legislation, neither poll measured the importance of this issue to voters in a context of competing public policy concerns. Environmental groups do not mention another national environmental poll conducted by Pollara in May 1998, which indicated that species at risk failed to register in the top 10 among Canadians' 'most important environmental issues' (Pollara, 1998), that recent polls eliciting more generally respondents' perception of the 'most important problem' in Canada consistently find that the environment does not make the top 10. In recent years, environmental concerns have played second fiddle to such issues as unemployment, national unity, and deficit reduction.

This finding is consistent with the distribution of costs and benefits of species-at-risk protection. The benefits of measures to protect species at risk are widely shared by all for whom such species have 'existence value'. However, the costs of protecting those species tend to be concentrated on a smaller number of individuals who are precluded from pursuing economic opportunities, such as harvesting trees and developing their land. When alerted by a polling firm to a policy proposal from which they could potentially benefit, the electorate at large registers their support. But given how small the average Canadian's stake is in species-at-risk protection, voters understandably tend to devote their political attention and energy to other, higher-priority issues. In contrast, those who have a lot to lose should species-at-risk legislation become a reality are more attentive and politically engaged in their opposition (Olson, 1965).

This characterization of public opinion helps to explain the failure of Bill C-65 and, more generally, the absence of national species-at-risk legislation in Canada. Environmentalists have been disadvantaged by the fact that, while Canadian voters may care in theory about the issue of protecting species at risk, in practice they tend to be ignorant about species-at-risk law and preoccupied with other policy issues. The Canadian government is undoubtedly cognizant that while enactment of tough species-at-risk legislation can offer only limited political support from a distracted electorate, it has the potential to generate strong and enduring opposition from those who would pay the price.

THE IMPACT OF IDEAS

The treatment of ideas as a causal variable has gained popularity throughout the political science community over the past decade (Hasenclever et al., 1996; Hall, 1997; Jacobsen, 1997). It is useful to view the impact of causal knowledge on species-at-risk policy through the lens of Haas's concept of epistemic communities (Haas, 1992; Adler and Haas, 1992; Haas, 1993). Haas defines an epistemic community as 'a network of professionals with recognized expertise and competence in a particular domain and an authoritative claim to policy relevant knowledge within that domain or issue area' (Haas, 1992: 3). In explaining international co-operation on environmental policy under conditions of seemingly conflicting interests, Haas emphasizes the influential role of scientists who have reached consensus on both the nature of the problem and policy prescriptions, particularly when those scientists infiltrate national and international bureaucracies.

A strong case can be made that an epistemic community exists within the species-at-risk policy arena. Both in terms of causal and normative beliefs about species endangerment, the consensus among conservation scientists is nothing short of remarkable. Dubbed a 'mission-oriented' scholarly endeavour, conservation science is an unabashedly value-laden enterprise committed to preservation of biodiversity

(Soule, 1985). In terms of causal ideas, there is virtually unanimous scientific consensus that habitat loss is the most important problem facing Canada's species at risk. Conservation scientists have translated that shared causal belief into an equally consensual policy recommendation that the protection of habitat must be the central focus of any legislation (Noss and Cooperrider, 1994; Noss et al., 1997).

Underscoring the unanimity of the scientific community on species at risk is the quite extraordinary political mobilization of Canadian scientists on this issue. A letter signed by over 300 scientists was sent to Sheila Copps in October 1995, criticizing the lack of focus on habitat issues in the discussion paper. This was followed by a letter to Copps's successor, Sergio Marchi, in February 1997, which was highly critical of Bill C-65. Thereafter, over 600 scientists signed a letter addressed to Prime Minister Chrétien in February 1999, demanding federal legislative action after the failure of Bill C-65.[1] Their message was blunt: 'Can you assure us that your government's new bill will include mandatory protection for all scientifically listed species, and nationwide habitat protection? Anything less will be scientifically unacceptable' (Abrahams et al., 1999). Bolstered by support from 13 fellows of the Royal Society, well-known environmental advocates such as David Suzuki, and even a dozen or so provincial and federal government scientists, Canada's scientific community employed the strength of their scientific consensus and their public credibility to promote their shared policy objectives.

Assessing the impact of this community is another matter, however. Haas's emphasis on the influence of consensual epistemic communities suggests that conservation scientists should have had a significant impact on the federal legislative process. Yet both the continued absence of federal species-at-risk legislation and the particular language of the failed Bill C-65, especially its weak provisions for habitat protection, call into question the influence of causal knowledge in Canadian policy. However, the possibility remains that the scientific community did contribute to the demise of C-65. The impact of ideas is best understood in terms of their role in fusing the campaigns of the scientific and environmental communities. The two communities' shared interest in maximizing habitat protection provisions allowed each to benefit from a symbiotic alliance, with environmentalists offering valuable organizational support and scientists in turn lending their credibility to the campaign against Bill C-65. In this sense, the failure, if not the content, of C-65 is consistent with the hypothesized influence of ideas.

ACTORS, INTERESTS, AND STRATEGIES

Discussions emphasizing the role and influence of North American interest groups in public policy formation are chiefly situated within the dominant framework of pluralism. Although traditional pluralist theory conceived of interest groups as competing on a relatively level playing field (Lindblom, 1959), this assumption has been reformulated in light of the contributions of theorists such as Wilson (1975), Lowi (1964), and especially Olson (1965), who stress how the concentrations of costs and benefits of public policy can promote or discourage collective political action. An

'Olsonian' analysis of species-at-risk protection would suggest that the diffuse rewards of preservation present obstacles to collective action by the potential beneficiaries, while the potential for concentrated costs provide ample motivation for political action by those in the business community who would bear the costs of habitat protection measures. The analysis would thus predict a biased interest group competition, with business interests tending to prevail over environmental interests.

There is no question that interest group conflict over the scope and substance of federal species-at-risk legislation is fundamental in explaining the failure of Bill C-65. Two main camps of actors can be discerned: an environmental camp comprising environmental groups and the politically engaged scientific community; and an opposing camp comprising industry groups and landowners fearful of restrictions on their property rights. It is important, however, to note the distinct position of Aboriginal groups, who did not fit neatly in either the environmentalist or business camp. First Nations argued that they should have been consulted in the legislative development process as a third order of government, not simply treated as another 'stakeholder', and that their wildlife management authority pursuant to land claims agreements should have been duly recognized. Yet, since the latter concern was largely addressed by amendments made by the committee, we conclude that the failure of Bill C-65 was not a function of Aboriginal groups' dissatisfaction.[2] The discussion that follows thus focuses on opposition to the bill from the environmental and business communities.

Strategies and Positions of the Environmental Community

Although Canada's environmental groups have often been described as 'fragmented' (Wilson, 1992), the issue of species-at-risk legislation prompted the creation in March 1994 of a peak association, the Canadian Endangered Species Coalition (CESC), directed by a steering committee of six major environmental groups and supported by over 100 other national and regional organizations extending well beyond the environmental community (Elgie, 1995).[3] CESC's key strategy has been to demonstrate (and promote) pan-Canadian solidarity for species-at-risk protection. They sponsored letter-writing campaigns from a diverse array of 'communities', including Canadian artists, US environmental groups, and, as noted above, scientists.[4] Indeed, according to one signatory to the scientists' letters, the Sierra Legal Defence Fund, which was a CESC steering committee member, effectively orchestrated the scientists' letter-writing campaign.[5] Sierra Legal Defence Fund staff maintain that they provided only 'administrative assistance' in distributing the petition among the scientific community, mailing the final version to Prime Minister Chrétien, and helping to correct legal facts within the letter itself. However, even this administrative role demonstrates a high level of interaction between the two communities.

Several features of Bill C-65 failed to satisfy legislative goals of the environmental community (CESC, 1997a). Along with the scientific community, environmentalists were critical of sections 13 and 14 of the bill, which provided for COSEWIC 'recommendations' regarding species listing but left the final say over listing of species at risk to the federal cabinet. Environmentalists and scientists were unequivocal in their denunciation of what they perceived as an opportunity for political interference in matters of science.[6]

An even greater deficiency of the bill from the environmental community's perspective was the absence of a habitat-based approach to species protection. Scientists and environmentalists alike were dumbfounded by the federal government's use of the unscientific term 'residence' to describe the area that would be protected by the federal government. Section 2 of the bill defined 'residence' as 'a specific dwelling place, such as a den, nest or other similar area habitually occupied by an individual during all or part of its life cycle' (Government of Canada, 1996: 4). This left other critical living areas, such as mating grounds, feeding grounds, and migration routes, unprotected. Although the Standing Committee amended the definition of 'residence' to include 'breeding and rearing areas', environmentalists and scientists still failed to secure any automatic prohibitions against the destruction of critical habitat comparable to those found in the US Endangered Species Act.

Another concern raised by environmental groups was Bill C-65's limited scope of application. Environmentalists recognized that co-operation among jurisdictions would be a crucial component of any habitat protection regime. However, they tended to offer a more expansive interpretation of the federal government's environmental jurisdiction than other players in the policy community, including the federal government itself. Environmentalists' desire for a more proactive federal presence was driven by their long-standing concerns over provincial 'conflicts of interest', especially in regard to the regulation of extractive resource industries such as mining and forestry (Harrison, 1996: 125). In particular, environmentalists were concerned that the National Accord is not binding on the provinces and therefore cannot guarantee a Canada-wide safety net for species protection.

The environmental community was quick to condemn Bill C-65 as a highly flawed safety net. In particular, environmentalists were critical that, because the bill offered full federal protection only to species living on 'federal lands', aquatic species falling under the Fisheries Act, and migratory birds covered under the Migratory Birds Convention Act, the bill did not apply throughout Canada, covering less than 4 per cent of the country's land base south of the territories (Austen, 1997).

Another difficulty environmentalists had with the bill's scope concerned the provisions for international and provincial transboundary species. Bill C-65 offered no protection for species at risk that cross interprovincial borders. With regard to the international border, section 33 as introduced by cabinet would have granted the federal government discretionary authority to protect internationally ranging or migrating animals and their residences. Although environmentalists were pleased that this was upgraded by the Standing Committee to *require* such protection, unless the federal government determined that a province had equivalent regulations, environmentalists wanted the section to be extended to include plants as well (Austen, 1997). However, provincial governments, industry, and private landowner interests had a very different view of section 33, depicting it as a Trojan horse through which the federal government could construct an overbearing federal role in regulating economic activity throughout Canada.

In some ways, environmental groups were pleasantly surprised by the federal government's proposals for species-at-risk recovery. Bill C-65 *required* the development of recovery plans within a year after the listing of an endangered species, and within two years for threatened and extirpated species. Vulnerable species were to be managed according to a plan developed within three years of listing. However, environmentalists were critical of the government's decision to leave *implementation* of such plans to the discretion of the responsible minister. In effect, Bill C-65 required development of recovery plans, but did not mandate their implementation.

Strategies and Interests of Industry and Private Landowner Groups

Perhaps the most important difference between the Canadian and US experiences in enacting federal species-at-risk legislation has been the presence of opposition from private landowners and industry groups. When the US Endangered Species Act was adopted in 1973, endangered species protection was considered a technical matter. Business and industry groups did not perceive the legislation to be a threat to their interests and did not even bother to testify during the House and Senate hearings (Yaffee, 1982). Indeed, endangered species protection was so politically popular that Republicans and Democrats in Congress practically stumbled over each other to champion the cause of bald eagles (Yaffee, 1982, 1994; Kohm, 1991). However, the controversial history of the ESA's implementation has belied the assumption that species protection is costless, a fact not lost on Canadian agricultural and industry groups. With the effective shutdown in the early 1990s of logging in the Pacific Northwest to protect the endangered spotted owl fresh in their minds, Canadian private-sector interests mobilized to oppose comparable legislation in Canada.[7]

Although their interests in species-at-risk policy were quite similar, and are thus analysed jointly here, the participation of industry and private landowner groups in debates over Bill C-65 was not co-ordinated. The interests of private landowners in opposing restrictions on land use were most effectively presented by the National Agriculture Environment Committee (NAEC).[8] Cognizant of environmental restrictions on land use faced by their counterparts south of the border[9] and still angry over the federal government's gun control legislation, farmers and ranchers were especially sensitive to a perceived federal 'attack' on individual and property rights. Various industry sectors were also outspoken in the debate over C-65. Although at least some industries have an interest in maintaining and promoting a 'green' image, particularly in the wake of international boycotts of Canadian forest products, they also have a tangible interest in protecting their investments from federal regulatory constraints.

Industry and landowner concerns about Bill C-65 centred on three issues: compensation, scope of application, and civil suits. First and foremost, private-sector actors argued that the goal of species preservation must be balanced against socio-economic realities. In short, they did not want to bear the financial brunt of species protection, arguing aggressively for a 'shared-cost' system. Their concern was that the federal government had not demonstrated its willingness 'to put the money where its environmental mouth had led it'.[10] Although section 8 of the bill authorized discretionary expenditures on wildlife conservation programs, it was not reinforced by the government's policy statements nor had there been any promises of compensation for future costs of regulatory compliance.

Second, industry and landowner groups were wary of any increased federal 'intervention' into what they considered to be provincial jurisdiction. This issue was particularly salient for industry groups, which were united in their opposition to a regime of overlapping federal and provincial regulations concerning species at risk. Although such groups emphasized the potential costs of regulatory duplication, inconsistency, and uncertainty, they invariably advocated the provinces take the lead role. As Harrison (1996: 176) has noted elsewhere, 'It is no accident that industries facing . . . regulations historically have favoured provincial jurisdiction, since they benefit not only from a symbiotic relationship between resource owner and developer, but from the threat of jurisdictional mobility.' Industry and landowners were particularly strongly opposed to the section 33 provisions for federal intervention in cases of transboundary species. Given unofficial Canadian Wildlife Service reports that suggest that such species could account for over 90 per cent of Canada's listed animals (Aniskowicz, 1998),[11] industry and landowner groups' unequivocal opposition to this provision is understandable.

Finally, compounding the lack of compensation measures and the international transboundary species provisions for the private sector were Bill C-65's non-discretionary provisions backed by the threat of civil suits. As depicted by industry and agricultural groups, the section 60 provisions for 'endangered species protection action' would only serve to engender unnecessary conflict among environmental groups, rural farming and industry communities, and the different orders of government. According to these groups, the proposed legislation undermined incentives for partnership and co-operation by concentrating too much attention on punitive measures and litigation.[12] It was also argued that a punitive regulatory approach would only invite intentional (yet virtually undetectable) elimination of endangered species on private lands—the 'shoot, shovel, and shut up' approach that some argue prevails under the non-discretionary American ESA (Andrews, 1996).

Summary
In the end, the proposed federal species-at-risk legislation foundered largely because it was opposed by all of the key actors. Although from the scientists' and environmentalists' perspective Bill C-65 was too weak a proposal to begin with, at the same time it contained elements that threatened industry and landowners. The Liberal government was thus unable to craft a minimum winning coalition of support and was forced into a politically embarrassing retreat.

It is noteworthy, however, from the summary in Table 7.1 that the interests of the key non-governmental actors were not entirely incompatible. Clearly, environmentalists' demand for non-discretionary language backed by citizen suits and the private sector's opposition to a regulatory approach, especially when combined with citizen suits, are fundamentally at odds, as are environmentalists' preference for a strong federal role and the preference of many business groups for provincial leadership. However, neither environmentalists' demands for both an independent listing process and a habitat-based approach to species protection and recovery nor business and landowners' call for compensation was opposed by the other camp. The bottom line is that environmentalists want a high level of habitat protection for species at risk throughout Canada, and industry and agricultural interests don't want to pay for that. However, if *someone else* was going to pay, the business community could conceivably be willing to concede to at least some of environmentalists' demands. In fact, both the multi-party Task Force on Endangered Species Conservation (1996) and the subsequent Species at Risk Working Group were able to achieve consensus on independent listing, habitat-based management, and compensation for regulatory 'takings'. There is also considerable potential for alignment, albeit along a different axis, of business and provincial interests in light of many business groups' opposition to regulatory intervention, especially by the federal government. Such areas of potential compatibility are significant in that they open the door for coalition-building in future efforts to craft species-at-risk legislation.

At the end of the day, it is difficult to ascertain which stakeholder groups 'won' or 'lost' this particular legislative battle since both camps opposed the bill. At one level, the failure of Bill C-65 can be viewed as a victory for industry and private landowner interests since these groups are favoured by continuation of the status quo—no regulation being better than some regulation. This is consistent with the Olsonian logic presented above. However, that analysis fails to acknowledge uncertainty with respect to the final outcome. If the federal government tries again and eventually passes an environmentally stronger bill than C-65, the ultimate winners will be the environmental community. This question will be revisited in the conclusion.

INSTITUTIONS

Having outlined the dynamics of interest group participation, we now examine how Canada's federal system of governance has influenced the development of species-at-risk policy. Weaver and Rockman (1993) have theorized that major policy initiatives are more difficult to enact in the horizontally fragmented institutions of the US congressional regime than in parliamentary systems. However, Canada's vertical fragmentation of authority may also create significant obstacles to policy change. Hoberg has suggested that 'Canadian federalism may prove to be a greater obstacle to environmental policy than the separation of powers in the United States' (1997: 378). The main effect of fragmentation, vertical or horizontal, is that it increases the level of political support necessary to achieve policy change (Banting, 1987). While its Westminster-style cabinet government concentrates a great deal of power at each level of government, in areas of overlapping jurisdiction like the environment Canada's decentralized federalism creates formidable obstacles to action. To explain the absence of Canadian federal legislation with the scope and stringency of the ESA, one must look to the allocation of constitutional authority and the relationship between the federal and provincial governments.

Due to a lack of explicit constitutional provision, the Canadian environmental policy field is characterized by overlapping and often unclear federal and provincial jurisdiction. Within this context of jurisdictional uncertainty the political struggle over federal species-at-risk legislation has been waged. There is no question that provincial governments have extensive constitutional authority to protect species at risk by virtue of their proprietary authority over Crown lands and their equally extensive authority to regulate matters concerning 'property and civil rights' and natural resources within the province. However, as in other areas of Canadian environmental policy, the extent of federal jurisdiction is less clear.

Although the federal government maintains proprietary authority over federal property within the provinces, these lands are limited in area in comparison to those controlled by the provinces.[13] The federal government can easily justify a legislative presence in wildlife matters that fall under its constitutional authority over fisheries, lands reserved for Indians, and those migratory bird species covered by the 1916 Canada/US Migratory Birds Convention. However, the cumulative potential of these heads of power is not sufficient to justify a sweeping federal species-at-risk statute. Nonetheless, a number of legal scholars, including a recently retired Supreme Court justice, have argued for a more expansive interpretation of the federal government's species-at-risk jurisdiction (La Forest and Gibson, 1999; Canadian Bar Association, 1996). In their view, the federal residual power to enact laws for the 'peace, order, and good government of Canada' could support federal legislation over all international and interprovincial transboundary species. The criminal law power is also relevant in light of a recent Supreme Court decision that upheld the federal government's authority to regulate toxic substances nationwide.[14]

Accepting that the functions and responsibilities of the federal and provincial governments concerning species at risk are not divided into watertight jurisdictional compartments, federal and provincial governments have resorted to the informal institutions of 'executive federalism' (Smiley, 1987). The forum for intergovernmental negotiation regarding species at risk has been the Wildlife Ministers Council of Canada (WMCC). The consensual nature of deliberations in the WMCC are noteworthy. By requiring that unanimity be achieved among ministers, the Council effectively gives each province a veto over any joint federal-provincial vision of Canada's species-at-risk protection policies.

Provincial sensitivities concerning the federal role in species-at-risk protection echo the intergovernmental debates over previous federal environmental statutes, including the Canadian Environmental Protection Act (CEPA) and the Canadian Environmental Assessment Act (Harrison, 1996). At one level, provincial ministers and bureaucrats personally resent the assertion of a federal leadership role in a field that has long been dominated by the provinces. However, at a deeper level, assertions of federal environmental jurisdiction, including authority to protect species at risk, represent a more fundamental threat not just to provincial environmental protection efforts but to provincial authority to direct economic development within their borders, particularly through exploitation of their Crown lands. This was brought home by a series of landmark court decisions culminating in the Supreme Court's Oldman River Dam decision, in which the federal government was forced by environmental groups to undertake belated environmental assessments of provincially owned projects (ibid.).

The provinces drew two lessons from that experience. The first was the need to avoid a legislative regime that could grant Ottawa a veto over either the provincial governments' own projects or private projects they had already approved. The second concerned the balance of authority between federal and provincial governments and citizens. The litigation over environmental assessment had demonstrated that non-discretionary language in a law or regulation could allow citizen groups to force the federal government to undertake actions that neither it nor the provinces desired. Citizen suits had the potential to move policy decisions from the private venue of intergovernmental councils to the open and less predictable forum of the courts, where the provinces had virtually no influence over the outcome.

These lessons prompted the provinces to pursue two complementary strategies. The first was to replace federal unilateralism with multilateral federal-provincial decision-making. 'National' or 'Canada-wide' standards, to be established by consensus among the federal government and the provinces, were offered as an alternative to 'federal' standards. The second strategy was to advocate reduction in overlap via a 'single-window' approach. As proposed by the provinces, provincial governments would be responsible for developing and implementing regulatory strategies to pursue agreed-upon national objectives.

The provincial governments felt they had accomplished these objectives in the National Accord for the Protection of Species at Risk, which was supported in principle by the federal government and all provinces but Quebec on 2 October 1996.[15] The Accord was an intergovernmental pledge to enact a seamless web of federal and provincial statutes and to provide for intergovernmental co-operation in their implementation. To promote this co-operative approach, the National Accord proposed to established a new intergovernmental body called the Canadian Endangered Species Conservation Council (CESCC) to be composed of the federal, provincial, and territorial ministers responsible for wildlife.

It should be emphasized that federal-provincial negotiations concerning species-at-risk protection were taking place within the context of a larger effort to restore intergovernmental harmony and rationalize federal and provincial roles in the environmental field. The Canadian Council of Ministers of the Environment, a parallel council to the WMCC with many overlapping members, was engaged in a 'harmonization' initiative dedicated to eliminating overlap and duplication between federal and provincial environmental programs. In early 1996 the first ministers had directed their environment ministers to produce an agreement by the end of the year (Harrison, 2000a). The result was approval in principle of the Canada-Wide Accord on Environmental Harmonization in November 1996, one month after the National Accord for the Protection of Species at Risk was signed. The Canada-Wide Accord is quite consistent with the provincial objectives noted above, proffering 'Canada-wide' national standards as an alternative to 'federal' ones and single-window implementation, with the anticipation that in most cases the window would be a provincial one.

Having supported the National Accord in principle, the provinces, it can be assumed, supported the enactment of federal species-at-risk legislation, at least in principle. Even the most adamant supporters of provincial paramountcy over species at risk recognized the need for a federal statute concerning federal lands and species, provided that it did not interfere with their role as primary landowners and resource managers. Nonetheless, the proposed Bill C-65 was roundly criticized by the provinces in this context of intergovernmental harmonization via decentralization. Alberta and Quebec in particular were quick to object to Bill C-65 on the grounds that it overstepped the boundaries of federal environmental authority and smacked of the confrontational American ESA. However, the criticism was not limited to these two provinces—Bill C-65 was rejected by every provincial government.[16]

The provinces were opposed to three provisions in particular in Bill C-65. First, they objected to section 13(2), which stipulated that COSEWIC members would be appointed by the minister after consultation with the provinces, but this did not *require* provincial consent. A second, and more serious, objection concerned the proposed federal government role with respect to international transboundary species. Echoing industry and private landowner criticisms, the provincial governments argued that the international transboundary species provisions granted Ottawa too much regulatory power, even though it applied only to animals and their 'residences'. Since most of Canada's wildlife species at risk range across the Canada-US border, the fear was that section 33 'could, quite conceivably, apply to every species in the province with the possible exception of some obscure frog in the northern Highlands.'[17] When this provision was upgraded by the Standing Committee to *require* federal involvement unless there were equivalent provincial regulations in place, the provinces objected to the prospect of 'having their performance judged by the federal government'.[18]

Finally, linked to these international transboundary issues were provincial arguments against the inclusion of a civil suit provision. Especially in light of the Standing Committee's amendments to render section 33 non-discretionary, the provinces did not relish the thought of environmental groups forcing the federal government to restrict activities on provincial lands to protect transboundary animal species. There was simply too much potential for federal interference in provincial economic development strategies and too little opportunity to work out an intergovernmental compromise when the venue shifts from the WMCC to the courts.

The question remains why, after years of negotiations with the provinces on species-at-risk protection and just a month after reaching agreement on the National

Accord, the federal government proposed a bill that so offended the provinces. It should be noted that the National Accord provides a framework for Canada's species-at-risk protection regime, but does not set out a clear division of jurisdictional authority. Thus, although the federal government may have intended to craft a bill that satisfied the provinces by largely confining the federal role to federal lands and species, they underestimated provincial sensitivities to assertion of a federal role with respect to transboundary species, sensitivities that were only exacerbated by the Standing Committee's amendments.

Provincial governments wanted federal legislation that would help to cement a more decentralized vision of Canadian environmental responsibilities; instead, they felt they were being asked to accept a regime of federal leadership and provincial junior partnership. Provincial governments thus added their voices to the chorus of objections to Bill C-65. That their objections were raised at a time when first ministers had made environmental harmonization a priority undoubtedly enhanced their impact.

It would be a mistake, however, to view the intergovernmental conflict over species protection exclusively as an institution-driven turf war. The provinces are not only governments, but landowners as well. In many respects, their objections to Bill C-65 paralleled those raised by private-sector interests, who noted that the price of protection of species at risk is often paid by those on whose land those species exist. Moreover, for the provinces, the legislative debate was not merely about protection of species at risk, but more broadly about control over economic development. Their electoral fortunes are more often determined by the jobs they create by promoting economic development and exploiting Crown resources than by the species they save. The compatible position of private landowners, industry groups, and provincial governments, therefore, is an interest-driven by-product of the institutional arrangements governing Canada.

CONCLUSION

On the face of it, the failure of Bill C-65, the Canadian Endangered Species Protection Act, is quite easy to explain—everyone hated it! As environmentalists Rita Morbia and Elizabeth May (1998: 19) observed, 'The bill was an odd combination of being weak and belligerent, ineffectual yet bullying.' Environmentalists felt it was too weak to protect species at risk effectively, scientists argued that the bill did not reflect scientific consensus concerning the centrality of habitat to species at risk, private-sector interests objected to the costs that might be imposed on them under the bill's regulatory approach, and the provinces objected to the potential for federal interference in their promotion and regulation of economic activity within their borders. In this sense, the outcome is consistent with all three elements of the regime: the bill did not reflect the state of ideas in the form of scientific knowledge, organized actors in the policy community all opposed it, and Canada's federal institutions invited influential provincial opposition. Interaction among the regime elements is also evident. The ideas put forth by the scientific community reinforced the position of environmental groups. And the institutional fragmentation of Canada's federal system provided an additional outlet for private-sector concerns through provincial governments.

However, such an analysis belies obvious tensions between some of these factors. At one level of analysis, the failure of C-65, which guaranteed that the status quo of no federal legislation would prevail for at least a few more years, can be viewed as a victory for the provinces and business. In that sense, the bill's demise is consistent with an Olsonian analysis of interest group influence and with the hypothesized effect of federal institutions, but inconsistent with the influence of epistemic communities as purveyors of causal ideas.

A comparison of the failure of Bill C-65 and the continuing absence of federal species-at-risk legislation with the adoption of the US Endangered Species Act in 1973 is consistent with this analysis, pointing to the influence especially of institutions and business interests. The resistance in Canada to the non-discretionary language and citizen suit provisions that are the hallmark of the ESA reflects the influence of Canada's parliamentary system of government with its fusion of the executive and legislative functions (Harrison and Hoberg, 1994). While a US Congress distrustful of the executive branch has incentives to include action-forcing language and citizen checks on implementation in the legislation it writes, a Canadian cabinet that both introduces and implements legislation has no incentive to adopt comparable measures that would fetter its own discretion. Differences between the US and Canadian federal systems are also relevant (Harrison, 2000b). With more extensive ownership of public lands, broader environmental jurisdiction via the interstate commerce clause, and more and thus individually weaker US states than Canadian provinces, it is unquestionably easier for the US than the Canadian federal government to assert a leadership role relative to subnational governments.

With respect to actors, a critical difference between the fate of Canada's Bill C-65 in 1997 and the US ESA in 1973 was the presence of opposition from the business community in Canada. It is noteworthy, however, that *ideas* help to explain the very different balance of interest group forces in the two cases. Yet the ideas at issue are not the scientific understanding of threats to endangered species protection discussed above, but rather greater awareness of the political and economic consequences of species-at-risk protection.[19] In effect, the Canadian business community learned important lessons from observing the impacts of protection measures south of the border.

Considered at another level, however, who ultimately wins, and thus which regime components prevail, depends on what happens in the next round. If the government ultimately adopts species-at-risk legislation that is stronger than Bill C-65, the failure of the first bill will represent a victory for environmentalists and the scientific community over business and the provinces,[20] an outcome that would be consistent with Haas's theory of epistemic communities and with an interest group analysis that lends greater credence to unselfish motives than does the Olsonian view.[21]

EPILOGUE: THE POLITICS OF THE SPECIES AT RISK ACT

The extent of opposition to Bill C-65 from all sides clearly indicates the magnitude of challenge faced by the federal government in the second round of species-at-risk policy-making.* However, in the late 1990s, three factors offered at least the potential to tip the scales in favour of compromise. First, as noted above, the compatibility (or at least absence of conflict) between the positions of various actors on some issues created the potential for political alliances. In fact, a more conciliatory dynamic among non-governmental stakeholders emerged in the aftermath of Bill C-65, when a quite remarkable constellation of environmental, industry, and private landowner groups came together in April 1998 to form the ad hoc Species at Risk Working Group (SARWG, 1998).²² This group subsequently achieved consensus on a proposal for a habitat-based approach to wildlife management and recovery of listed species, so long as it included compensation from federal and provincial governments for private landowners, business, and community participants. An important development in background conditions also emerged in the wake of the failure of Bill C-65: a change in the fiscal climate. At the end of the decade, the era of deficit-driven budget cuts ended and federal purse strings had been loosened somewhat. A costly compensation package was thus no longer inconceivable. Finally, growing threats posed by environmentally conscious foreign consumers of Canadian products, especially forest products, appeared to have increased support from the business community for *federal* species-at-risk legislation, which could serve to bolster the reputation of Canadian exports.

The stage thus was set for another attempt at compromise between the disparate actors in the policy community. Moreover, the federal government's appointment in August 1999 of David Anderson, a senior minister with environmental credentials, to the Environment portfolio as a replacement for Christine Stewart, a junior minister widely criticized by environmentalists, signalled its commitment to move forward. In December 1999, the federal minister released an update outlining the government's new three-pronged strategy: federal-provincial co-operation; co-operative, voluntary stewardship programs backed by federal funding; and regulatory legislation (Environment Canada, 1999). Bill C-33, the Species at Risk Act (SARA), was subsequently introduced in the House of Commons in April 2000.

The new bill departed from its predecessor in three important respects. First, the new bill represented a shift in the choice of central policy instrument from regulation to spending, which offered at least the potential to shift the politics of Canadian endangered species protection from making enemies to making friends (Hoberg and Harrison, 1994). Although the proposed legislation still provided for extensive regulatory authority, both the discussion paper and the minister himself stressed that voluntary, publicly funded stewardship programs would be relied on in the first instance, with resort to mandatory restrictions only if '*all possible* stewardship, incentive and other efforts are insufficient' (Environment Canada, 1999: 16, emphasis added; also

* This section of the chapter was written only by Kathryn Harrison and George Hoberg, who accept full responsibility for the analysis as well as any errors of fact or interpretation. The section was written without the assistance of William Amos.

McIlroy, 1999b). Even then, the bill offered the prospect of compensation to the private sector under conditions to be set out in companion regulations to the legislation. To reinforce these programs, the federal government's year 2000 budget committed $90 million over three years and $45 million per year thereafter to programs to protect species at risk.

The second change concerned the scope of federal jurisdiction asserted. Emboldened by the Supreme Court's recent decision to uphold CEPA based on the federal criminal law power,[23] the federal government claimed authority to protect not only transboundary species as in Bill C-65, but to protect *all* species at risk *anywhere* in Canada, including on private and provincially owned lands. (Transboundary species other than migratory birds were no longer even mentioned.) This approach also facilitated greater emphasis on habitat than in Bill C-65, with the federal government now claiming authority to protect individuals, their residences,[24] and 'critical habitat' of threatened or endangered species anywhere in Canada. This represented an important extension of Bill C-65, which focused on the residence of at-risk species and the species themselves on federal lands only, with the exception of 'federal species' (e.g., aquatic species, migratory birds, and transboundary species).

A third change of approach was clearly intended to counterbalance the second, however. The federal government's greater jurisdictional assertiveness in theory was accompanied by greater deference to provincial governments in practice. There was a clear presumption of federal deference if the provinces were doing the job. Thus, the automatic prohibition against harming listed species and their residences applied only on federal lands, except in the case of aquatic species and migratory birds, which would have been protected anywhere in Canada. Although Bill C-33 would have authorized cabinet to apply these prohibitions elsewhere, that was left to cabinet discretion. Similarly, the assertion of federal authority to protect critical habitat anywhere in Canada would have been applied in the first instance only to critical habitat on federal lands designated by the minister, and only to other lands if the minister recommended to cabinet that a province or territory had failed to provide adequate habitat protection. This deferential approach with respect to critical habitat applied even to international migratory bird species, for which federal constitutional authority is quite clear.

This discretionary and residual federal role was a departure from both the equivalency approach proposed by the Standing Committee with respect to transboundary species and the equivalency provisions of the 1987 Canadian Environmental Protection Act, in which a federal regulatory role is asserted in the first instance, and withdrawal is authorized only if a province can satisfy conditions demonstrating that it is implementing equivalent policies. The bill thus had more in common with residual federal environmental legislation of the 1970s than the more jurisdictionally assertive CEPA and Canadian Environmental Assessment Act of the last decade (Harrison, 1996). In effect, the safety net was broader but further below the action.

Each of these three key departures from Bill C-65 can be seen as a response to objections raised by one of the three key groups: business and private landowners, the provinces, and environmentalists. It would appear that the federal government had attempted to fashion a compromise that would draw at least some support from each camp.[25] From the perspective of the private sector, the proposed reliance on regulation via the criminal law power was significantly softened by the priority given to voluntary and publicly funded stewardship activities. Accordingly, the government's communication strategy emphasized 'balance' and the need for a 'co-operative Canadian approach' as an alternative to the litigation of the US Endangered Species Act (see, e.g., Duffy, 2000a, 2000b). This was further reinforced by the prospect of compensation for regulatory takings and the elimination of the civil suit provisions of C-65, which had been a red flag to business and landowners cognizant of the impacts of environmentalists' lawsuits concerning the spotted owl on the US forest industry. In shifting from a regulatory to an expenditure-based approach, the federal government apparently sought to buy off stakeholder opposition with the dollars of inattentive taxpayers.

In response to provincial objections to threats of 'federal intrusion' in Bill C-65, Bill C-33 offered stronger guarantees that the provinces and territories, which were now treated for all intents and purposes like provinces, would be the lead actors. Automatic protections for international transboundary species and their residences were withdrawn. While the assertion of broader federal authority to protect critical habitat of these and other species throughout Canada still represented a significant threat to provincial autonomy, that threat was softened by provisions restricting federal involvement unless provinces and territories invited it or failed to do the job.

A key critique of Bill C-65 offered by environmentalists was that it only applied to federal lands, constituting 4 per cent of the country south of the territories (though they tended, perhaps disingenuously, not to emphasize the more far-reaching potential for federal involvement to protect transboundary species). In response, SARA promised a broader federal safety net that could apply to any threatened or endangered species anywhere in Canada. Moreover, authority to protect habitat, another central concern of environmentalists and their scientific allies, was significantly extended. Finally, the promise of federal funding for stewardship activities was not only consistent with the co-operative approach advocated by many in the environmental community, but it is noteworthy that some conservation groups could have benefited from funding for their own stewardship activities.

The 2000 election was called just after the committee hearings on Bill C-33. As a result, it is impossible to say whether these efforts to forge a minimum winning coalition were successful. The prospects for success did seem better in this round. Industry and private landowners had either been quiet or offered mildly supportive comments on the bill. The positions of these groups would have been clarified during committee hearings, but the election was called before that happened. Because the debate over the thorny issue of compensation was avoided, it is difficult to determine the level of support in the private sector. For many groups, future support will undoubtedly turn on the terms and generosity of compensation.

Despite the broad assertion of federal jurisdiction under criminal law, most provincial and territorial governments also were surprisingly quiet. While none offered enthusiastic public expressions of support for Bill C-33, only Alberta and Quebec questioned the constitutionality of the legislation, a contrast to the unanimous provincial opposition to Bill C-65 (Teel, 2000; Francoeur, 2000).

The effort to appease environmentalists was considerably less successful, however. Conservation groups such as Ducks Unlimited, the Nature Conservancy, and the BC Wildlife Federation, all of which stand to gain federal financial support for their own stewardship activities, have been supportive of the government's new emphasis on stewardship and have been willing to participate in press conferences announcing federal funding initiatives. They did not, however, mount public campaigns in favour of the bill (one exception is the BC Wildlife Federation). In contrast, virtually all other major environmental groups, led by the Canadian Endangered Species Campaign, the Sierra Legal Defence Fund, World Wildlife Fund Canada, and the Sierra Club, opposed the proposed legislation. It would appear that the promise of greater habitat protection and a broader safety net do not outweigh those groups' opposition to a decentralized and discretionary approach. Canadian environmentalists have historically been distrustful of provincial governments as resource owners and proponents of economic development (Harrison, 1996), and have also favoured federal legislation because it offers an opportunity to fight battles once at the national level rather than 13 times, in each province and territory. Although environmentalists had supported co-operation via funding for stewardship programs and compensation for regulatory takings, the bill further reinforced this co-operative approach by withdrawing the civil suit provisions and extending discretionary language, changes that were strongly opposed by most Canadian environmentalists. Finally, environmentalists achieved no gains in their fight against 'political listing' of species at risk by cabinet rather than independent COSEWIC scientists having the final word on this matter.

In crafting the Species at Risk Act, the federal government may have attempted to fashion a compromise with something for everyone: deference to the provinces, greater recognition for First Nations, compensation and stewardship as an alternative to regulation and citizen suits for business and landowners, and a broader federal safety net for environmentalists. However, in practice the bill can be seen as a shift in the direction of a provincial-business coalition of support, offering both a more deferential and discretionary federal role, with expenditures as a substitute for regulation. If the bill is reintroduced and passes in its current form, this outcome would seem consistent with the impacts of federal institutions and interest group competition biased in favour of business, as hypothesized above. However, it remains to be seen both whether the government has enough support from the provinces and business community and whether it is possible for a Canadian government to fashion a minimum winning coalition for national environmental legislation without significant support from the environmental community. If they cannot, and the next attempt to enact federal endangered species legislation also fails, it seems unlikely that any future Canadian government will wade into the perilous waters of endangered species politics for a very long time.

Notes

1. The letter-writing campaigns have been given ample exposure through Canada's newspapers. For a sample of this coverage, see Hanna (1997), McIlroy (1997), and McIlroy (1999a).
2. This conclusion was reinforced during interviews with Environment Canada officials.
3. Supporters of CESC include the Canadian Bar Association, the Canadian Labour Congress, Greenpeace, the United Church of Canada, the National Farmers' Union, the Body Shop of Canada, the Council of Canadians, United Fish and Allied Workers Union, United Steel Workers of America, and Vegetarians of Alberta. The Coalition underwent administrative reforms in early 1999 and was renamed the Canadian Endangered Species Campaign.
4. Canadian artists, including Margaret Atwood, Robert Bateman, and Bruce Cockburn, were organized through CESC to write a joint letter to Sergio Marchi in the spring of 1996, just prior to the release of Bill C-65. US environmental groups were mobilized in a similar fashion, adding an international element to the environmental community's approach.
5. Confidential interview with scientist, July 1999.
6. As the scientists' letter to Prime Minister Chrétien opined: 'Canada's endangered species are too imperiled, too close to extinction and too precious to be held hostage to lobbyists, political manipulation or simple ignorance' (McIlroy, 1999a: A1).
7. Elsewhere, Hoberg (1991: 110) has argued that policy convergence can occur through 'activist-driven emulation', whereby knowledge is transmitted through 'transnational policy communities' as actors try to 'shame' their government to enact policies similar to those in other countries. It is interesting to note that in this case a dynamic of 'reverse emulation' is occurring, in which Canadian landowners are attempting to steer the government away from a statute like the ESA and are thus advocating policy divergence (Amos, 1999).
8. The NAEC is a peak association representing the vast majority of Canada's agricultural producers on environmental issues, including the Canadian Federation of Agriculture, the Canadian Pork Council, the Dairy Farmers of Canada, and the Western Canadian Wheat Growers Association.
9. In addition to controversies over endangered species, groundwater protection measures have been highly controversial in recent years in the United States. BC farmers were also familiar with tight restrictions on land use as a result of that province's Agricultural Land Reserve.
10. Confidential interview, 16 July 1999.
11. 'Listed' here refers to all endangered, threatened, and vulnerable animals on the COSEWIC list.
12. An inflammatory article published in *Alberta Report* captured this sentiment: 'Endangered Species Overkill: Ottawa's proposed wildlife grab threatens property owners with huge fines, years in jail and loss of land' (Avram, 1997: 6).
13. 'Federal lands' included those in the Arctic, national parks and National Wildlife Areas, defence bases, Transport Canada properties, Native reserves, and federal Crown corporation lands.
14. *R. v. Hydro-Québec*, [1997] 3 S.C.R. 213.
15. Quebec did not sign the National Accord. However, interviews with provincial and federal wildlife officials indicate that Quebec was in agreement with respect to the principles of this framework agreement.

16. Confidential telephone interviews with BC, Ontario, and Quebec officials, July-August 1999.
17. Confidential telephone interview, July 1999.
18. Confidential telephone interview with Quebec wildlife official, July 1999.
19. Indeed, had the causal ideas advanced by the scientific community been an important influence, Canada would have stronger endangered species legislation than the US since, if anything, the scientific community has become even more aware of the importance of threats to habitat in the intervening years.
20. This also explains how environmentalists could claim victory when the bill failed, even though in the interim the failure of Bill C-65 resulted in a continuing situation of no national legislation.
21. To further complicate matters, what cabinet was collectively thinking when they sacrificed Bill C-65 (and thus, which factors influenced that round) may not be reflected in what direction cabinet ultimately takes in future endangered species legislation if background conditions or, as noted below, the constellation of interest group positions should change.
22. SARWG is composed of representatives from the Canadian Nature Federation, Canadian Wildlife Federation, Sierra Club of Canada, Canadian Pulp and Paper Association, Mining Association of Canada, and the National Agriculture Environment Committee. It is important to note, however, that the group met as individuals without any official mandate, and not as representatives of specific sectors.
23. *R. v. Hydro-Québec*, [1997] 3 S.C.R. 213.
24. The government also accepted the broader definition of 'residence' offered by the House of Commons committee considering Bill C-65, which includes breeding and rearing grounds.
25. It is noteworthy that concessions were also made in response to concerns raised by First Nations. SARA thus provided recognition in the preamble of the role of Aboriginal peoples in protection of species at risk, required consultations with First Nations in various circumstances, and recognized the primary role to be played by wildlife management boards under land claims agreements, akin to that of provincial and territorial governments.

REFERENCES

Abrahams, M., et al. 1999. Letter Addressed to the Right Honourable Jean Chrétien. <http://www.zoology.ubc.ca/~otto/LTR-ENG.htm>. Accessed 6 June 1999.

Adler, E., and P. Haas. 1992. 'Conclusion: Epistemic Communities, World Order, and the Creation of a Reflective Research Program', *International Organization* 46, 1: 367–90.

Amos, William A. 1999. 'Federal Endangered Species Legislation in Canada: Explaining the Lack of a Policy Outcome', MA thesis, University of British Columbia.

Andrews, Anthony. 1996. House of Commons Standing Committee on the Environment and Sustainable Development. Evidence, Meeting No. 50, 21 Nov. 1996, line #0.0845. <http://www.parl.gc.ca/committees352/sust/evidence/50_96-11-21/sust-50-cover-e.html>. Accessed 10 Aug. 1999.

Angus Reid Group. 1995. *Public Support for Endangered Species Legislation*. Toronto: Angus Reid Group.

Aniskowicz, Theresa. 1998. 'Cross-Border Species Listed by COSEWIC', unpublished document. Ottawa: Canadian Wildlife Service.

Austen, C. 1997. 'Evaluating the Proposed Canadian Endangered Species Act'. <http://www.umich.edu/~esupdate/library/97.03-04/austen.html>. Accessed 7 July 1999.

Avram, J. 1997. 'Endangered Species Overkill', *Alberta Report* 24 (24 Feb.): 6–9.

Banting, Keith. 1987. *The Welfare State and Canadian Federalism*, 2nd edn. Montreal and Kingston: McGill-Queen's University Press.

Canadian Bar Association. 1996. Letter Addressed to the Honourable Sergio Marchi and the Honourable Allan Rock, Re: Endangered Species Legislation. Ottawa: Canadian Bar Association.

Canadian Endangered Species Coalition. 1997. Federal Legislation Backgrounder: Summer 1997. <http://www.chebucto.ns.ca/Environment/FNSN/cesc/bck-su97.html>. Accessed 3 July 1999.

Canadian Wildlife Service. 1993. *Proceedings of Focus Group on Management of Species at Risk: Do We Have the Right Tools?* Ottawa: Environment Canada.

Cashore, Benjamin, George Hoberg, Michael Howlett, Jeremy Rayner, and Jeremy Wilson. 2001. *In Search of Sustainability: British Columbia Forest Policy in the 1990s.* Vancouver: University of British Columbia Press.

Doern, G. Bruce, and Tom Conway. 1995. *The Greening of Canada.* Toronto: University of Toronto Press.

Duffy, Andrew. 1999. 'Land owners crucial in conservation plan', *Ottawa Citizen*, 18 Dec.

——. 2000a. 'Anderson blasts U.S. critics', *Ottawa Citizen*, 4 Mar.

——. 2000b. 'Wildlife bill strikes balance, Minister says: Species at Risk Act a titanic failure environmentalists argue', *Ottawa Citizen*, 12 Apr.

Elgie, S. 1995. *Endangered Species Legislation: A Bear Necessity.* Toronto: Sierra Legal Defence Fund.

Environment Canada. 1994. *Endangered Species Legislation in Canada: A Discussion Paper.* Ottawa: Environment Canada.

——. 1995. *The Canadian Endangered Species Protection Act: A Legislative Proposal.* Ottawa: Environment Canada.

——. 1999. 'Canada's Plan for Protecting Species at Risk: An Update', Dec.

Fafard, Patrick, and Kathryn Harrison, eds. 2000. *Managing the Environmental Union: Intergovernmental Relations and Environmental Policy in Canada.* Kingston: School of Policy Studies, Queen's University.

Francoeur, Louis-Gilles. 2000. 'Espèces menacées: Québec accuse Ottawa de jouer dans sa cour', *Le Devoir*, 12 Apr., A3.

Government of Canada. 1996. Bill C-65 An Act respecting the protection of wildlife species in Canada from extirpation and extinction. Second Session, Thirty-fifth Parliament, 45 Elizabeth II.

Haas, Peter. 1992. 'Introduction: Epistemic Communities and International Policy Coordination', *International Organization* 46, 1: 1–36.

——. 1993. 'Epistemic Communities and the Dynamics of International Environmental Co-operation', in V. Rittberger, ed., *Regime Theory and International Relations.* Toronto: Oxford University Press, 168–201.

Hall, Peter. 1997. 'The Role of Interests, Institutions, and Ideas in the Comparative Political Economy of the Industrialized Economies', in M. Lichbach and A. Zuckerman, eds,

Comparative Politics: Rationality, Culture, and Structure. Cambridge: Cambridge University Press, 174–207.

Hanna, D. 1997. 'Scientists attack animal-protection act', *Vancouver Sun*, 5 Feb., A4.

Harrison, Kathryn. 1996. *Passing the Buck*. Vancouver: University of British Columbia Press.

——. 2000a. 'Intergovernmental Relations and Environmental Policy: Concepts and Context', in Fafard and Harrison (2000).

——. 2000 b. 'The Origins of National Standards: Comparing Federal Government Involvement in Environmental Policy in Canada and the United States', in Fafard and Harrison (2000).

—— and George Hoberg. 1994. *Risk, Science, and Politics*. Montreal and Kingston: McGill-Queen's University Press.

Hasenclaver, A., P. Mayer, and V. Rittberger. 1996. 'Interests, Power, Knowledge: The Study of International Regimes', *Mershon International Studies Review* 40: 177–228.

Hoberg, G. 1991. 'Sleeping with an Elephant: The American Influence on Canadian Environmental Regulation', *Journal of Public Policy* 2, 1: 107–32.

——. 1993. 'Environmental Policy: Alternative Styles', in M. Atkinson, ed., *Governing Canada: Alternative Policy Styles*. Toronto: Harcourt Brace Jovanovich, 307–42.

——. 1997. 'Governing the Environment: Comparing Canada and the United States', in Keith Banting, George Hoberg, and Richard Simeon, eds, *Degrees of Freedom: Canada and the United States in a Changing World*. Montreal and Kingston: McGill-Queen's University Press, 341–85.

—— and K. Harrison. 1994. 'It's Not Easy Being Green: The Politics of Canada's Green Plan', *Canadian Public Policy* 20 (June): 119–37.

—— and E. Morawski. 1997. 'Policy Change Through Sector Intersection: Aboriginal and Forest Policy in Clayoquot Sound', *Canadian Public Administration* 40, 3: 387–414.

International Fund for Animal Welfare. 1999. 'New poll shows overwhelming support for endangered species legislation', <http://www.ifaw.org/press/pr052699.html>. Accessed 30 May 1999.

Jacobsen, John Kurt. 1995. 'Much Ado About Ideas: The Cognitive Factor in Economic Policy', *World Politics* 47 (Jan.): 283–310.

Kohm, K. 1991. *Balancing on the Brink of Extinction*. Washington: Island Press.

La Forest, Gerard V., and Dale Gibson. 1999. 'Constitutional Authority for Federal Protection of Migratory Birds, Other Cross-Border Species, and Their Habitat in Endangered Species Legislation', unpublished paper, Nov. 1999.

Lindblom, Charles. 1959. 'The Science of Muddling Through', *Public Administration Review* 14: 79–88.

Lowi, T. 1964. 'American Business, Public Policy, Case Studies, and Political Theory', *World Politics* 16: 677.

McIlroy, A. 1997. 'Scientists fear bill won't save animals', *Globe and Mail*, 5 Feb., A6.

——. 1999a. 'Tough endangered species law demanded: 631 scientists tell Chrétien he must do more to save the animals', *Globe and Mail*, 24 Feb., A1, A11.

——. 1999b. 'Incentives proposed to protect animals', *Globe and Mail*, 17 Dec.

Morbia, R., and E. May. 1998. 'Unlikely Allies Join to Protect Canada's Species at Risk', *Global Biodiversity* 8, 3: 18–21.

Noss, R., and A.Y. Cooperrider. 1994. *Saving Nature's Legacy*. Washington: Island Press.

——, M. O'Connell, and D. Murphy. 1997. *The Science of Conservation Planning*. Washington: Island Press.

Office of the Minister of the Environment. 1996. Press Conference Notes, The Honourable Sergio Marchi, on the occasion of the tabling of the Canada Endangered Species Protection Act. Ottawa: Environment Canada.

Olson, Mancur. 1965. *The Logic of Collective Action*. Cambridge, Mass.: Harvard University Press.

Pollara. 1998. Canadians' Views on Climate Change. <http://www.pollara.ca/new/Library/Climate/Intro.html>. Accessed 3 July 1999.

——. 1999. *Canadian Attitudes and Opinions Toward Endangered Species—A POLLARA Report*. Toronto: Pollara.

Soule, M. 1985. 'What is Conservation Biology?', *BioScience* 35: 727–34.

Smiley, Donald. 1987. *The Federal Condition in Canada*. Toronto: McGraw-Hill Ryerson.

Species at Risk Working Group. 1998. *Conserving Species at Risk and Vulnerable Ecosystems: Proposals for Legislation and Programs*. Species at Risk Working Group.

Task Force on Endangered Species Conservation. 1996. *Task Force Report on Federal Endangered Species Legislation, Second Report*. 23 May.

Teel, Gina. 2000. 'Alberta opposes wildlife legislation', *Calgary Herald*, 12 Apr., A10.

United Nations. 1992. *Convention on Biological Diversity, Text and Annexes*. Montreal: Secretariat of the Convention on Biological Diversity.

Weaver, R.K., and B.A. Rockman. 1993. 'Assessing the Effects of Institutions', in Weaver and Rockman, eds, *Do Institutions Matter? Government Capabilities in the United States and Abroad*. Washington: Brookings Institution.

Wilson, J. 1992. 'Green Lobbies: Pressure Groups and Environmental Policy', in Robert Boardman, ed., *Canadian Environmental Policy: Ecosystems, Politics, and Process*. Toronto: Oxford University Press.

Wilson, J.Q. 1975. 'The Politics of Regulation', in J. McKie, ed., *Social Responsibility and the Business Predicament*. Washington: Brookings Institution.

Yaffee, S.L. 1982. *Prohibitive Policy*. Cambridge, Mass.: MIT Press.

Risk Politics in Western States: Canadian Species in Comparative Perspective

ROBERT BOARDMAN

The utilitarian dimension of biodiversity is at the centre of much environmental discourse on sustainability. For many wild species, however, no significant practical uses for human societies have been developed or even envisaged. Advocacy on behalf of such species is often linked to arguments about use, as in assessments of the benefits to local economies of national parks and their wildlife, or the value to agriculture, forestry, and pharmacological research of wild plant species. Arguments about the protection of turtles, marmots, and orchids, however, are not primarily practical-use arguments. What factors shape the responses of industrial societies to these questions? What are the main issues that arise in efforts to foster good programs of stewardship?

In this chapter I will draw on the experiences of a number of Western industrialized nations, particularly Australia, the European Union (EU) and its member states, and the United States. Through study of these, I argue, we can gain a valuable comparative perspective on endangered-species debates and politics within Canada, for example, the role of legislative frameworks in promoting conservation goals, approaches to questions of private landownership and habitat protection, and issues arising from multi-jurisdictional contexts. Setting Canada in the context of other OECD (Organization for Economic Co-operation and Development) countries gives us insights into the ways nations with broadly similar principles of economic organization, constitutional frameworks and democratic systems, and cultural values deal with species at risk. The differences among these countries, and the variability of policy responses, also shed light on these problems. Following a brief overview of the threats to species, I will focus on four areas—the roles of NGOs, the capacities of governments, international pressures, and landownership issues—as core sets of problems in OECD countries. The emphasis is on broadly political matters, as decisions about threatened species of wild flora and fauna reflect significant choices by societies among environmental, land-use, and other values.

ENDANGERED SPECIES IN THE OECD COUNTRIES

The ecosystems of the Western industrialized nations vary widely. So do the nature and magnitude of the threats to wildlife populations on their territories. Among

Table 8.1: OECD Data on Threats to Wildlife

	Mammals Threatened		Birds Threatened		Fish Threatened		Reptiles Threatened		Amphibians Threatened		Invertebrates Threatened		Vascular Plants Threatened	
	Number	%	Number	%	Number	%	Number	%	Number	%	Number	%	Number	%
Canada	37	19.2	46	10.8	65	6.4	14	33.3	9	21.4	5	-	104	2.5
Mexico	163	33.2	178	16.9	120	5.7	127	18.0	49	16.9	32	0.1	446	2.5
US	49	10.5	79	7.2	64	2.4	26	7.1	8	3.6	-	-	118	0.5
Japan	14	7.7	54	8.3	22	11.1	20	20.4	15	23.4	118	0.4	1,870	26.7
Korea	17	17.0	59	15.0	12	1.3	3	11.5	2	13.3	43	0.3	58	1.5
Australia	47	14.9	50	6.4	17	0.4	51	6.6	29	14.3	-	-	1,085	4.3
N. Zealand	7	15.2	43	25.3	8	0.8	11	18.0	1	25.0	14	0.1	119	5.0
Austria	29	35.4	81	37.0	38	65.5	14	87.5	21	100.0	2,291	-	1,157	39.2
Belgium	18	31.6	46	27.5	25	54.3	2	50.0	4	30.8	373	39.7	383	31.9
Czech Rep.	30	33.3	123	66.1	19	29.2	11	100.0	21	100.0	165	0.4	1,100	43.7
Denmark	12	24.0	18	10.6	6	18.2	-	-	4	28.6	498	13.2	117	9.8
Finland	7	11.9	16	6.7	7	11.9	1	20.0	1	20.0	158	0.6	88	4.8
France	24	20.2	51	14.3	28	6.6	6	16.7	11	29.7	110	0.3	387	8.1
Germany	29	36.7	70	29.2	45	68.2	11	78.6	12	57.1	-	-	772	25.7
Greece	44	37.9	55	13.0	26	24.3	4	6.8	-	-	-	-	177	3.1
Hungary	59	71.1	70	18.8	26	32.1	16	100.0	16	100.0	>400	>0.9	495	19.8
Iceland	-	-	10	13.3	-	-	-	-	-	-	7	0.6	37	7.6
Ireland	2	6.5	42	21.8	9	33.3	1	33.3	1	33.3	-	-	9	0.7
Italy	38	32.2	117	24.7	-	-	13	22.4	9	23.7	2,435	4.3	270	4.8
Luxembourg	33	54.1	65	50.0	13	38.2	6	100.0	13	100.0	-	-	153	14.5
Netherlands	10	15.6	46	27.1	23	82.1	6	85.7	9	56.3	-	-	486	34.9

Table 8.1 — continued

	Mammals Threatened		Birds Threatened		Fish Threatened		Reptiles Threatened		Amphibians Threatened		Invertebrates Threatened		Vascular Plants Threatened	
	Number	%	Number	%	Number	%	Number	%	Number	%	Number	%	Number	%
Norway	3	5.9	14	6.3	-	-	1	20.0	3	50.0	-	-	87	7.3
Poland	13	15.5	39	16.6	13	27.1	3	33.3	18	100.0	4,864	16.5	226	9.8
Portugal	17	17.3	43	13.7	8	18.6	3	8.8	-	-	-	-	255	8.2
Spain	25	21.2	52	14.1	20	29.4	11	19.6	4	16.0	391	1.6	509	6.4
Sweden	12	18.2	21	8.6	7	12.7	-	-	7	53.8	711	3.0	214	11.3
Switzerland	27	34.2	84	42.6	21	44.7	11	78.6	16	94.1	839	37.3	579	22.1
Turkey	30	22.2	30	6.7	19	9.9	17	16.0	3	13.6	-	-	237	7.7
UK	14	22.2	35	6.8	6	11.1	3	42.9	2	28.6	976	4.3	204	9.1

SOURCE: OECD, *OECD Environmental Data: Compendium 1999* (Paris: OECD, 1999). Copyright OECD 1999. Percentages are in relation to the numbers of species known. 'Threatened' species are the totals in the IUCN categories 'critically endangered', 'endangered', and 'vulnerable'. There are some variations among OECD countries in the use and definition of categories of species at risk.

OECD countries we find remarkable ecological diversity. There is a wide range of biomes within individual countries such as the US, Australia, and Japan. Threats to wildlife species, and the problems facing governments and conservation organizations, are thus different from country to country.

In Australia, federal endangered species legislation was developed in 1992. The estimates at that time were that 43 mammal species were endangered or vulnerable (17 per cent of the total), together with 26 bird species (3 per cent); 209 vascular plants (1 per cent of the total) were endangered and 784 (4 per cent) vulnerable; 260 invertebrate species were threatened; and 9 species of amphibians, 20 reptile species, and 13 freshwater fish species were either endangered or vulnerable (Endangered Species Advisory Committee, 1992: s. 5). By the end of 1997 a total of 1,125 endangered and vulnerable species had been listed (Nadeau, 1999: 5). In many European countries, more than 45 per cent of vertebrate species have been estimated to be threatened. Over one-third of bird species have declining populations (EEA, 1998: ch. 8). Taking Western industrialized nations as a whole, OECD data indicate a widespread pattern of threats to wildlife (Table 8.1).

Species at risk are not distributed uniformly, even within the same country. Threats to habitat and species in Canada are most pronounced in southern regions. Studies from other countries suggest that threatened or endangered populations tend to be concentrated in relatively few areas. In the US most endangered species are clustered in 'hot spots' in southern California, the southeast coastal states, and the southern Appalachian region (Dobson et al., 1997: 550–1). As well, there is variation over time. In the two centuries of European settlement, one-half of Australia's forests have been cleared for agricultural and other uses. There has been major ecological degradation in more than one-half of the country's arid and semi-arid areas. Half of the world's species of mammals that have become extinct over this period have been in Australia. An estimated 100 species of vascular plants have become extinct there since the late 1700s (Bates, 1995: 300). Much habitat destruction took place in Europe before the rise of conservation movements, as in the large-scale coastal wetlands drainage schemes in the Netherlands in the seventeenth century for agriculture and, in England, the destruction of the water meadows of southern Lancashire for industrial development and urban expansion in the late nineteenth century.

Several common factors threaten species and habitats in Western countries. Spreading urbanization, industrial development, and intensive agriculture have had widespread consequences throughout Europe. Many butterfly species are threatened or endangered as a result of loss of habitat and the use of pesticides in agriculture. In the Netherlands problems are intensified by high population density and high levels of urbanization and agricultural development. Its location around the estuaries of the Rhine and Meuse rivers means that the Netherlands also has to cope with river-borne pollutants from other countries (Van der Zande and Wolters, 1997: 219; Blenkinsop, 1999). There was a net loss of 53,000 miles of the traditionally species-rich hedgerows in the United Kingdom between 1984 and 1990, largely because of misguided management practices by farmers (This Common Inheritance, 1992: 74). Similar complexes of factors have left large footprints on Australia. For example, the subtropical Big Scrub rain forest, originally covering 75,000 hectares, had been reduced to 300 hectares and 10 remnant patches by 1900 (Endangered Species Advisory Committee, 1992: s. 5).

In the US, factors contributing to ecosystem deterioration, as in the fragmentation of secondary forests in Ohio, include clear-cutting, highways, gas pipelines, and strip mines. Dams, river diversion schemes, pollution, and other factors have had cumulative impacts on the ecological integrity of aquatic ecosystems (National Biological Service, 1995: 10, 13; USGS, 1999). Decades of canal- and dike-construction projects and related activities in the interests of agribusiness, urban development, and tourism have transformed much of Florida. This has focused attention on its endangered species, including recently the Cape Sable sparrow. It has been estimated that road networks and traffic have direct ecological effects—including the effects of noise on some bird species—on 22 per cent of the land area of the US (excluding Alaska and Hawaii) (Forman, 2000: 33–4).

The changing status of some wildlife species continues to be shaped by interactions with agricultural, hunting, and consumer economies and the transborder movement of people and goods. Species traditionally considered a threat by farmers include the Arctic fox in Iceland, the prairie dog in the US, and the lynx in Sweden. The hunting of migratory and other wild birds is a historic part of the informal economies of several regions of France and Italy. The 1990s saw a significant expansion in the use of alligator hides for clothing and other consumer products in the US, with a more than fourfold increase over the 1987–95 period (Colman, 1997). Problems related to introduced species have been a chronic feature of the Australian and New Zealand economies since the start of European settlement. In New Zealand, government programs in relation to some introduced species, particularly trout and the koire rat, have generated conflict with Maori principles of land management (Chanwai and Richardson, 1998). The rapid growth in the 1990s of North American ruddy duck populations in Europe, following escapes from an English waterfowl reserve, has threatened several native species. Changing geographical distributions of species over several centuries, whether or not these have been affected by human activities, create practical and definitional difficulties in Europe about which species should be protected or reintroduced. Finally, the uncertain implications of climate change during the next several decades are likely to have consequences for many wildlife species in Western countries. Butterfly ecology, for example, is particularly sensitive to temperature changes, and the impacts of climate change have already been documented in European studies (Roy and Sparks, 2000: 415–16).

NGOs AND SPECIES POLITICS

Group activities have traditionally characterized the wildlife politics of OECD states. A dynamic civil society is an essential foundation for endangered species protection. Such groups, numbering into the thousands, are diverse. They include organizations focused on a population or species, families of species such as owls or whales, particular habitats, or larger clusterings such as the transnational networks of wild bird groups. Some nationally based organizations with broad conservation mandates, such as the Sierra Club in the US, have credentials as pioneer environmental groups with histories dating back to the 1890s. Others, including the World Wildlife Fund (WWF), now one of the major global NGOs active on endangered species questions, grew out of the environmental movement of the 1960s.[1] Many groups cultivate wide public memberships, while some aim to mobilize scientific and technical expertise. Groups with interests across several categories of environmental issues, such as Greenpeace and Friends of the Earth, vary in the attention paid to threatened fauna and flora; some groups, as in British Columbia and Queensland, tend to be drawn into habitat protection issues in cases where these have dominated larger environmental agendas.

Such groups are 'political' in both the broader and narrower senses. Even the smallest groups have the goal of educating people and raising awareness of threats to wildlife species and habitats. Often in coalitions or informal networks, many groups engage in activities that are directly, through lobbying or presentations to parliamentary committees, or indirectly aimed at influencing governments. Many organizations are autonomous actors with significant programs of their own. These activities include the purchase or management of critical habitats by nature trusts. Increasingly since the late 1980s, groups have co-operated with government agencies in a variety of partnership roles. The process has helped blur still further the lines between government and civil society. These organizations have a different character as we move from country to country.

Influencing Governments

The capacity of wildlife protection groups to influence the course of government policy depends on a number of factors and circumstances. Even within the same country, groups vary considerably in their lobbying skills and the size of their memberships and funding bases. Frequently, temporary alliances are formed to pursue common goals, as in the Canadian Endangered Species coalition. Groups in some countries have been able to foster links with political parties. A tradition of support for environmental issues on the part of the Australian Democrats, for example, has been a significant feature of environmental politics in Australia. Public attitudes may be resistant to conservation goals, and these views influence governments. For example, there is evidence that the grey wolf population in Minnesota was reclassified from endangered to threatened in 1978 primarily because of hostile public attitudes to wolves (Brown, 1997: 175).

There are also differing conceptions of governance on the part of groups. As in Canada, the older nature conservation organizations in Europe, the US, and Australia were often politically conservative. Quiet diplomacy, élite (and, in Europe, often aristocratic or royal) connections, and the goal of incremental change of laws tended to be the norm. This philosophy came under attack in the 1980s and 1990s. The Australian Conservation Foundation (ACF), for example, was transformed into a more activist organization. The move was accelerated and symbolized by the election of Peter Garrett, a lawyer and the lead singer of the Midnight Oil rock band, as president. The ACF, along with Greenpeace, the Wilderness Society, and other groups, took a leading role in 1999 in criticizing the federal government's proposed revisions to national environmental legislation, while WWF Australia, the Tasmanian Conservation Trust, and others found the political compromises underlying the changes broadly acceptable (MacDonald, 1999: 1).

The constitutional structures of states also affect the strategies and efficacy of groups. In 1987 Swiss environmental NGOs used that country's permissive referendum rules to secure a constitutional amendment on the protection of moors. The action was triggered by Defence Ministry plans to set up an army post in Schwyz canton. The campaign succeeded despite sustained opposition from the federal government and parties in parliament (Gottesmann, 1997: 217). Groups in several Western countries have become increasingly adept at using the courts to gain greater protection of species and habitats. Provisions of the bird and habitat directives of the EU have been used by groups to obtain court rulings designed to pressure national governments into compliance. However, the capacity of US conservation organizations to adopt this strategy has been diminishing in the 1990s, following more stringent definitions by courts of the rules allowing such actions (Glaberson, 1999). The US system, though, remains probably the most open of all OECD countries. Groups have relatively easy access to congressional hearings on environmental legislation. The constitutional separation of powers makes these a more pivotal part of national environmental debate and policy-making than tends to be the case with the work of parliamentary committees in Canada. The drawback, for environmentalists, is that Congress is also exposed to multiple pressures from other groups. Local interests often aim to derail measures to extend the reach of species protection. Conservation groups have accordingly lent broad support to more direct methods for protecting areas by the executive branch, in relation to public lands, that bypass Congress. The strategy was used by President Bill Clinton from 1993 in Arizona, Colorado, Utah, and other states.

Environmental groups operate in a pluralist context of competitive politics. Advocacy of protection of species and habitats is only one of the voices heard inside governments. The hunting of wild birds in rural areas of Italy and France re-emerged in the mid-1990s as one of the most heated arenas of environmental politics in the EU. French hunters' organizations became increasingly concerned at signs that key EU directives on species and habitat protection, as well as France's own wild bird protection laws, would eventually be enforced under pressure from the European Court of Justice. There has been a long history of expedient neglect by authorities, and thus a reluctance on the part of the police to enforce regulations. The French hunting constituency is the largest of any European country, with about 1.6 million registered hunters and 3.5 million occasional hunters. In 1998 a political party with the single-issue platform of defence of hunters' rights got 4 per cent of the French vote in elections to the European Parliament (*The Economist*, 1998).

In the US, the Endangered Species Act (ESA) became the subject of heated political controversy in the late 1970s. This was precipitated by events such as the temporary halting in 1978 by the courts of the Tennessee Valley Authority's Tellico dam project, which was found to be a threat to the habitat of a small fish, the snail darter.
Amendments made to the Act in 1978 were widely criticized by environmentalists as
a retreat from the principle of endangered species protection (Hartmann, 1981: 179-
86). Growing perceptions of the ability of endangered species legislation to restrain
economic activity led to the politicization of the ESA's species listing process amid
attacks from conservative critics and to its temporary halting by the Reagan administration in the early 1980s. The requirement to consider the economic impacts of listing or threat-abatement measures has become a feature of legislation and court interpretations across several jurisidictions in the US and Australia.

Civil Society Dynamics
Groups are important, too, as autonomous social actors. Seeing them primarily in
terms of their ability to influence governments misses much of the texture of species
conservation action. Some roles have grown naturally out of traditional concerns, for
example, the collection of good data on species at risk, the purchase or management
of critical or important habitats, and programs for species recovery.

Much of the data needed in relation to species distribution in Europe, and especially
in Britain, have been traditionally generated by volunteer recorders. Differences in collection formats, however, frequently make the task of integration of data for analysis and
policy difficult (Gaston et al., 1998: 362; Griffiths et al., 1999: 330). While some landowner groups in the US have been associated with activities that degrade habitats, others have
been active in species conservation. A group in Riverside County, California, for example, raised more than $25 million in the 1990s to fund ecological research on the kanga-
roo rat (Sheldon, 1998: 304). Social attitudes towards different species thus come into
play. It is often difficult for conservation groups to mount campaigns or recovery programs around amphibians and reptiles, spiders and other invertebrates, or non-flower-
ing plants. In Europe, as in Canada, conservation knowledge in these areas tends to be
correspondingly weaker than for many mammal and bird populations.

Collaboration between conservation groups and other organizations has become a
defining characteristic of much endangered species activity in OECD countries. Some
countries have experimented, as has Canada, with round-table formats for bringing
together groups with differing perspectives on conservation issues. Much of this work
is locally based, as a couple of recent examples from Scotland show. In the Outer
Hebrides, a coalition of local landowners, crofters, and environmentalists was formed
in the late 1990s to combat threats from minks to seabirds and other local species.
Mink populations have grown throughout Britain as a result of releases from fur
farms staged by animal liberation groups.[2] In a more ambitious scheme, the major
power company in Scotland was compelled in 1999, following a campaign by local
conservation groups, to establish and manage a habitat for two golden eagles threatened by a Strathclyde windpower project.

Relations between groups and governments have been shaped since the mid-1980s by the larger context of government reorganization and deficit reduction in OECD countries. Conservation groups have moved into some niches vacated by retreating bureaucracies. They are often welcomed as partners by government environmental agencies, which need tangible evidence of external constituency support to fight internal bureaucratic battles. They also increasingly rely on NGOs to pursue conservation objectives. In Britain, wildlife groups routinely receive matching funds or grant aid from the government. These organizations have become essential to the success of species recovery projects in which government agencies have a stake. Programs on toad conservation, for example, have developed as partnerships between English Nature, a government agency, and the Herpetological Conservation Trust, an NGO (Beebee and Denton, 1995: 26–7). There is a wide range of partnership undertakings across other EU countries, particularly since the adoption of the 1992 habitats directive. For example, programs of land purchase, education, and research on the part of several types of organizations have been central to the conservation of peatlands in Ireland since the mid-1980s (IPCC, 1997).

THE CAPACITIES OF GOVERNMENTS

Just what the governments of Western nations should be doing to protect species and habitats, and how well they do what they do, has been a highly contested issue. Policy options aired in public debates vary from advocacy of strong leadership roles by governments, grounded in effective legislation and wide-ranging regulatory power, to support for minimalist strategies of respecting property rights and steering clear of limiting private-sector economic activity. A wide spectrum of alternatives has been discussed in Canada. As in other federal systems, the role of the national government is constrained by the constitutional rules and conventions governing its relations with provincial governments. In Australia and the US, the states, like Canada's provinces, retain important constitutional powers in relation to environmental policy, natural resources, and land use. The 1990s saw a partial withdrawal of OECD governments from many policy areas they had moved into in earlier decades. There was growing public-sector and private-sector unease with the idea of government intervention generally, especially in its more regulatory forms. The capacities of governments to engineer change and oversee sound species and habitat protection programs vary considerably.

Endangered Species Regimes

The absence of national endangered species legislation in Canada has made it an anomaly among OECD countries. Of all these, the US has been home to the most ambitious, systematic, regulatory, and politically controversial endangered species legislation. Many of the legislative and administrative arrangements in the US for national parks, wildlife refuges, forest protection, wilderness, and other areas are much older than those on endangered species (Littell, 1992: ch. 2; Bean, 1983: chs 6–8). Federal law began in 1966 and culminated in the regime established by the ESA in 1973.

It is complemented by other measures such as those on marine mammals, bald and golden eagles, and migratory birds. Central to the ESA ethic from the outset was the assumption by the federal government, primarily through the Secretary of the Interior and the US Fish and Wildlife Service, of responsibility for listing species of flora and fauna facing various degrees of threat. Protection and compliance were to be secured through a system of penalties and other enforcement mechanisms. Several core aspects—particularly the focus on individual species rather than habitats or ecosystems and the nature of the incentives to landowners to manage properties well—were at the centre of political controversies during the 1980s and 1990s.

The Australian endangered species regime has been built in part on lessons learned through study of the US experience, modified to take account of the much more restricted manoeuvring room of the national government within Australia's federal system. Three overlapping phases are discernible.

The first, as a result of the National Parks and Wildlife Act and specific measures such as those on international trade in endangered species and protection of the Great Barrier Reef, saw the creation of a distinct conservation role for federal author-ities in the 1970s. Second, in 1992 the Endangered Species Protection Act gave Canberra authority to identify and list threatened wildlife species, promote recovery plans and projects aimed at threat abatement, and encourage landowners to take part in voluntary species and habitat protection agreements. The Act's three schedules were designed for the listing, respectively, of species, ecological communities, and threat processes. Australia's endangered species legislation had to deal with many of the controversial issues that surfaced later in Canadian debates. In the species listing process, for example, the minister has final authority, following consideration of advice given by the endangered species scientific subcommittee (sections 24[1] and 159[1] of the 1992 Act). Expanded federal funding for the recovery of endangered species and other biodiversity-related programs was later assured through the estab-lishment of the Natural Heritage Trust of Australia Reserve in 1997. The third and cur-rent phase was initiated with the Environment Protection and Biodiversity Conservation Act of 1999. This legislation formally put species and habitat questions in a larger environmental policy framework. A central aim of the legislation was to tilt the balance of environmental policy power back towards state governments. Federal advances in the environmental policy area date back to the early 1970s, but Canberra's powers have been continually vulnerable to attacks from the states. Under the provi-sions of the 1999 Act, the federal government's role was to be triggered by issues requiring a 'national' response. These issues were essentially in areas carved out by Canberra in previous legislation on endangered species, national heritage sites, nuclear issues, and Australia's responsibilities in international environmental law, for example, in relation to wetlands under the 1971 Ramsar Convention and the determi-nation of World Heritage Sites.

Western Europe has still more complex jurisdictional issues. In addition to their own endangered species laws and programs, the member states of the EU, currently 15, are co-participants in a regional European polity. Though not a federation, the EU constitutes in effect a European level of government. Its powers are currently defined in the Amsterdam treaty, which entered into force in 1999. The environment, including provisions for endangered species and protected areas, falls within the common pillar of EU policy-making. Decisions are made collectively by member states and are binding on each one. Policy processes include majority voting, or decisions by consensus, in the Council of Ministers. This body meets according to policy area, so that environmental matters are handled in it by the environment ministers of the national governments. Environmental policy authority is shared with the European Commission, the bureaucracy of the EU. This level of European governance interacts with national institutions and, through the principle of subsidiarity, with regional and other institutions.

Two measures in particular define the scope of EU authority in relation to species and habitats. The wild birds directive of 1979 obliges states to protect species listed in two annexes. Governments are required to set up special protection areas and to institute penalties for killing, removal of eggs, or habitat disturbance (Grossman, 1997: 29–30). The habitats directive of 1992 broadened the scope of these provisions. Unlike its predecessor, this was not primarily species-based. It also established stricter measures for the protection of threatened habitats. The goal is the eventual creation of a network of protected areas, Natura 2000, which will incorporate important ecological provisions such as buffer zones and wildlife corridors. National governments designate protected areas following criteria defined by the EU, and they are then obliged to institute protection measures (Diana, 1998; Bennett, 1994; Déjeant-Pons, 1998).

Unfortunately, the presence of legislation on species at risk does not necessarily mean that governments are effective in protecting them. There have historically been marked differences among OECD countries in the way nature conservation responsibilities are organized inside governments (Baldock, 1987). The political will to pursue conservation objectives varies, as does institutional capacity in terms of personnel and funding at the disposal of key agencies. Several departments or agencies are typically responsible, as is the case in the Canadian federal government, for closely related matters such as parks and protected areas, cultural and natural heritage sites, and the management of natural resources. More powerful departments often view wildlife either as an economic resource or as a nuisance or irrelevance. In Canada, Australia, and other parliamentary systems, these diverse voices are represented and compete in cabinet deliberations on environmental policy. Jurisdictions vary, too, in the attention paid to endangered species in environmental impact assessments related to government activities. The US Fish and Wildlife Service, for example, has not in practice been a barrier to government projects. Between 1987 and 1994 officials carried out nearly 100,000 assessments relating to the likely effects of federal agency projects on wildlife species, but these resulted in only 352 jeopardy opinions (Sheldon, 1998: 280).

The government agencies concerned with endangered species thus often have limited powers in practice to influence overall environmental policy. They are constrained in relation even to endangered species policy-making by pressures from other government departments and agencies. Environmental officials, moreover, particularly if they are located in an agency rather than a ministry headed by a cabinet member, in turn tend to be restricted in their capacity to shape other areas of government policy that have implications for environmental policy.

Federalism and Territorial Politics

A further complication for species-at-risk arrangements is that multiple territorial jurisdictions exist within countries. Federalism has the traditional virtue of encouraging geographical flexibility and responsiveness to local conditions and political cultures. However, it also makes for organizational complexity, regime variation, and slow process, and may weaken the prospects for effective national policy frameworks and regulatory standards. It is on the latter grounds, as well as in their criticisms of the inadequacies of specific provincial measures, that Canadian environmental groups have tended to argue in favour of a strong federal role in the protection of species at risk. The differing effects of territorial politics can be seen in the US, Australia, and the EU.

Differences between federal and state approaches to conservation have been a feature of US wildlife policies. In a recent study, Porter and Underwood (1999: 6–7) observed that federal national parks staff tend to be oriented towards minimizing human impacts on wildlife species, whereas the officials of state agencies are likely to be more concerned in practice with questions of wildlife use or, for example, in the case of deer, with regulating species abundance. Debates on endangered species in the US contain echoes of the pre-1970s regime, before Washington took over the 'fundamental responsibility' for protecting endangered species. Traditionally, Littell (1992: 95) writes, 'the regulation of wildlife was considered a state function, with no room for federal interference.' Washington is still limited in that its direct authority is restricted to federal lands. The extent of these varies in states. Hawaii, for example, has 225 ESA-listed species, but only 16 per cent of its land is federally managed; only 1 per cent of Texas, with 70 species, is federal land; by contrast, 68 per cent of Alaska is under federal jurisdiction, but that state has only five listed species (Sheldon, 1998: 285). In practice the two levels of government co-operate, and do so routinely on ESA listings. The participation of state wildlife and other agencies is also required for the implementation of federal conservation and recovery programs. In terms of personnel, state governments collectively outweigh federal capabilities. State regimes differ. Some go further than federal law requires. New York and California, for example, have been allowed by the courts to ban the trade in footwear made from wildlife skins not listed by the ESA (Littell, 1992: 96).

The states have traditionally been the key players on wildlife questions in Australia. Practices have continued to vary. Victoria took the lead among Australia's governments in adopting in 1988 the first law, the Flora and Fauna Guarantee Act, to be based on habitat or ecosystem principles as opposed to the US-style single-species approach (*Halsbury*, 1995: 180–9640).[3] The states, like Canada's provinces, are sensitive to federal interference. Their defensive rhetoric occasionally resorts to claims of the 'sovereign' powers of state governments. Contested definitions of Canberra's role were key elements in the explosive forest-protection politics of Tasmania and Queensland in the 1980s and 1990s. While Canberra received explicit powers for endangered species in the 1992 legislation, its direct authority was restricted to a relatively small territorial base of federal lands.[4] Federal powers beyond this are in practice those that can be negotiated with state authorities on the basis of legislative provisions. Mounting criticism of alleged federal intrusiveness led to a return to the principle of state primacy in the 1999 legislative restructuring. The political limits to Canberra's powers were illustrated by its secondary role in a major forestry deal negotiated in New South Wales in 1998. Environmentalists attacked the weak provisions for conservation targets in the agreement and the secrecy of the negotiations between state authorities and timber companies. The federal Minister of Conservation and Forestry, Wilson Tuckey, described the plan as laying the groundwork for the 'rape and pillage' of the state's northeastern forests (Finkel, 1998).

Territorial politics affecting endangered species are also prominent in Europe, both within the EU and outside. Switzerland, whose non-EU status appeared to be confirmed by the strong nationalist vote in its 1999 election, remains a heavily decentralized polity in which the cantons have the main responsibility for nature conservation and endangered species. Federal legislation dating from 1967 obliges the national government to protect important habitats and to protect species threatened with extinction (Gottesmann, 1997: 211–12). In practice, though, the national government in Bern tends to be a weaker player. Its role depends on the ways its powers to fund programs and to collect data on endangered species work out in its relations with individual cantons. Inter-cantonal arrangements are important. The Fanel wetlands, for example, are protected through co-operation between government agencies in Bern and Vaud cantons.

There are jurisdictional complexities, too, in non-federal systems in Europe. In Sweden, local governments are significant actors in national conservation networks. Recommendations for the establishment of protected areas normally originate in county administrations (Emneborg and Götmark, 2000: 729). In the UK, also formally a unitary state, a trend of growing decentralization of powers on endangered species questions was evident even before the creation in 1999 of parliaments for Scotland and Wales. This constitutional change followed the earlier breakup of the former central administrative agency for nature conservation into separate bodies for England, Scotland, and Wales. These tend to have differing approaches and programs on species recovery and habitat protection, which has prompted debate, at least in England, on the loss of a national voice on endangered species questions. As a result of constitutional change in the late 1970s Spain moved significantly away from the previous centralized regime, though without establishing a federal system. Nature conservation was one of the policy areas largely transferred to regional and other levels of government. One result of this decentralization has been competition among governments to create more and better protected areas. Some administrations, for example those of Andalusia and the Canary Islands, took an early lead in the designation of significant portions of their territory as protected (Morillo and Gómez-Campo, 2000: 166).

Evaluating Performance

Formal and informal monitoring of the efficacy and consequences of government actions surrounds endangered species programs. In Australia a 1997 review of progress since the enactment of the 1992 law called for speedier work towards recovery programs for listed species, higher levels of funding of endangered species programs, and greater attention to co-operation with landowners and private enterprise (Nadeau, 1999).

A common complaint in European assessments and monitoring by conservation groups is lack of compliance with EU measures by national governments. Although conservation goals are set at the European level, national governments are also vulnerable to counter-pressures by domestic groups. During the 1990s the European Court of Justice increasingly had a say in these conflicts. Key judgements in 1991 and 1993 against Germany and Spain underscored the obligations of governments to protect wild bird habitats. Yet, by the end of 1997 only two countries, Belgium and Denmark, were regarded as having a complete system in place (*Natura* 4, Nov. 1997: 9). The weaknesses of the wild birds directive have been described by Kiss and Shelton (1997: 206) as 'the most important problem of compliance among EC environmental measures'.

In American debates we similarly find both friends and foes of government species-at-risk programs. Both sides complain of politicization and the power of special interests, though from different angles. Critics sympathetic to the general goals of habitat protection have argued that more than three decades of federal measures have produced mixed or minimal results. The analysis by Loomis and Echohawk (1999) of the National Wilderness Preservation System indicates that of 35 American ecological regions, 23 have less than 1 per cent of their land area protected.[5] Further, little-known species are sometimes neglected by the ESA. In 1995, 53 listed species received less than $1,000 each for research and recovery activities, while over one-half of ESA funding was allocated to 10 species, including four species of Pacific salmon (Baker, 1999: 279).

Brown (1997: 171–2) is among many critics who have argued that the ESA has been ineffective in protecting threatened species. By 1997, an additional 948 species had been added to the 134 on the original ESA lists. Of the few species delisted, seven or eight delistings resulted from data errors (including the Tumamoc globeberry and McKittrick pennyroyal), and 10 or 11 were delisted because populations were considered restored. The latter group includes the American alligator, Palau owl, and Arctic peregrine falcon. Brown argues, however, that in most cases ESA actions cannot account for species recovery. Some bird populations recovered because of the DDT ban, for example, and species on the Palau islands in the western Pacific recovered following US troop withdrawals. In her view this leaves only one species, the American alligator, 'for which the ESA can claim sole credit'. Similarly, of reclassifications from endangered to threatened (20 species between 1973 and 1997), only a few improvements can be attributed to the ESA. The chief problem, Brown asserts, is that the ESA is 'biologically flawed' and an inadequate protector of habitats. Its approach is 'inadequate to protect ecosystems and biodiversity. The Act's protections are too little, too late. Recovery plans are not based on ecological factors, but are focused on high profile, popular species.'

By contrast, Smith and Smith (1997: 17, 20), writing about the lessons Canada can learn from the ESA, approach these data from the perspective of the alleged negative consequences of ESA meddling, particularly on grounds of property rights and bureaucratic neglect of the economic costs of conservation decisions. They suggest that the Tumamoc globeberry and other cases indicate that species listings 'are often far more motivated by the desire to halt projects on government and private land . . . than by any genuine concern about the species.' They also find fault with the biological concept of 'species' that underlies the legislation. This explains, in their view, 'why the number of listed species is so high and why the public has an inflated picture of the degree of endangerment.'

External Pressures on States

International agreements have become increasingly important influences on endangered species politics and practices (van Heijnsbergen, 1997: 22–42). Many of the arrangements by governments and NGOs for protection of birds in Canada stem from the Canada-US agreement on migratory birds of 1916. Multilateral environmental agreements and the environmental provisions of regional trade arrangements such as NAFTA create obligations for governments, even though much international law still lacks the enforcement mechanisms that apply in domestic law. Among other things, the 1992 Convention on Biological Diversity committed states adhering to it to more effective national protection of endangered species. States make commitments by signing and ratifying such agreements, but subsequent follow-up actions may be delayed or reshaped in the moulds of domestic politics. This has happened in Canada in relation to the 1997 Kyoto Protocol on climate change. Unanticipated consequences also follow international agreements. For example, provisions of the Kyoto Protocol create incentives for companies and states to engage in reforestation in order to acquire carbon credits. However, since the type of forest land is not specified, critics have argued that some projects, such as monoculture plantations, could be promoted that have harmful effects on biodiversity.

The efficacy of many of these arrangements is limited by traditional conceptions of sovereignty. The scope of the Convention on International Trade in Endangered Species (CITES) is formally restricted to species and their products that enter international trade. A significant gap in international environmental law could be filled by the initiation of a parallel convention on endangered species within countries, perhaps on the basis of the kinds of listing approaches found in the international agreements on wetlands protection (the Ramsar Convention) and World Heritage Sites (UNESCO's World Heritage Convention). This kind of step, however, would eat more deeply into national sovereignty than countries historically have found tolerable. The domestic political consequences of international instruments can nonetheless be far-reaching. The UN Biological Diversity Convention was a significant stimulus to conservation and species recovery action on the part of governments and NGOs. In the

UK it led directly or indirectly to the production of action plans for more than 100 species and 14 habitats by the end of 1995 (*This Common Inheritance*, 1996: 15).

This external dimension takes on special significance for federal countries. In Canada and other federal systems, governments normally have constitutional powers in relation to international agreements and international trade. Support for agreements such as CITES thus helps guard federal authorities against the threat of growing domestic marginalization—even though their ability to implement such agreements still depends on state or provincial governments. This delicate balancing act has been a feature of Austrian (Kiss and Shelton, 1997: 183) and Australian (Marlin, 1998) endangered species policies. In several Australian states, environmental NGOs have made use of the provisions of international agreements to urge Canberra to take a more proactive stance on habitat and species protection and to confront foot-dragging state governments from this higher moral ground.

The political attractions of international leadership roles, and the prestige that can flow from them, form a consideration for some Western governments. The Netherlands and Switzerland each have long historical records of initiatives on international nature conservation issues, as does Canada in the IUCN/World Conservation Union and other fora. Traces of this heritage can still be seen in the respective approaches of these countries to multilateral co-operation. Germany hosts the international secretariats set up through the agreements on climate change, migratory species, and desertification, and in 1999 it helped to create new facilities in Bonn for the Environmental Law Centre of the IUCN/World Conservation Union (*Common Ground*, 1999: 5). Notions of global and hemispheric leadership also shape US conceptions of international environmental policy, as in its role in the creation of CITES in 1973 and with initiatives in relation to such issues as Japanese whaling, the dolphin-tuna issue, and debt-for-nature swaps in Central America.

Finally, environmental politics are shaped in part by actors' recognition of the fact that transborder ecological processes affect the status of wildlife species. The hemispheric dimension of nature conservation has been an important feature of US policies and NGO programs for several decades. Canadian approaches to many areas of environmental policy have been defined in practice in a continental context, which, as in the cases of acid precipitation, migratory birds, and Great Lakes water quality, acknowledges that national actions alone cannot solve environmental problems. Van der Zande and Wolters (1997: 219–20) argue that nature in the Netherlands is 'part of a European and global ecological system'. Dutch nature conservation, they write, 'must therefore be embedded in an international framework'. The regional setting is likewise a characteristic dimension of endangered species and environmental policy generally in each of the Scandinavian countries (Christiansen, 1996).

Europe's experience with international cross-border protected area arrangements goes back to a 1925 agreement between Poland and Czechoslovakia. The earliest species agreements arose from hunting problems and date from the 1780s. The protection of several species rests on inter-state co-operative mechanisms, for example, between France and Italy for the ibex, and for seal conservation in the Waddenzee among Denmark, Germany, and the Netherlands. In all there are currently some 24 such arrangements involving a total of 20 countries (Kiss and Shelton, 1997: 197–8; van Heijnsbergen, 1997: 9–14).

LAND-USE POLITICS AND SPECIES AT RISK

Governments have important roles to play in the creation and management of national parks and protected areas, and in ensuring that species recovery and the maintenance of ecological integrity remain a crucial part of the responsibilities of these. The habitat of many species, however, is on private land. The negotiation of easements and other kinds of conservation agreements with landowners—including households, woodlot owners, farmers, and forestry and mining companies—is thus critical to the success of national protection efforts.

Such measures should be seen in the larger context of patterns of land use in OECD countries. Jurisdictions differ in the percentage and representativeness of national parks and other protected areas, and in regard to the rules and practices governing mining, agricultural development, highway construction, and other activities. Endangered species have a varying place in these frameworks. Before the 1960s, national parks in Canada and the US were viewed by governments more in terms of their recreational potential than as areas for protecting habitats and species and maintaining ecological integrity (Dubasak, 1990: 96–7). Competing economic interests are at stake in protected-area and species-recovery initiatives, just as they are in any form of land-use negotiations (Brown and Shogren, 1998).

The 1998 forestry agreement in New South Wales arose in part from timber companies' interest in acquiring land in order to obtain carbon credits in the post-Kyoto climate change regime. Despite extensive research and data-gathering, the final deal was relatively weak in terms of conservation targets for high-priority species. Land designated for conservation tended to consist of escarpment forest of little interest to the companies, and included areas already relatively well represented in protected-area planning. As a result of the agreement one species, the Hasting's River mouse, may become extinct as its remaining refuge is in fragmented areas in the northeast of the state (Finkel, 1998).

In the United States, more than 50 per cent of ESA-listed species have at least 81 per cent of their habitat on private land (Sheldon, 1998: 285). Endangered species legislation has thus unavoidably fostered a constituency of landowners worried about its implications for property rights. The matter became increasingly politicized from the late 1970s. The Reagan administration's 1982 amendments to the ESA tried to promote greater sensitivity to the interests and concerns of landowners. Disputes since then have been frequent. Some have centred on the federal government itself as a landowner. In 1997, the US Navy finally agreed to take steps to protect the San Clemente loggerhead shrike, an endangered species on an island owned by it since the 1930s, under threat of court action by the American Bird Conservancy (Line, 1997).

The 1990s saw much greater use of habitat conservation plans negotiated with companies under the ESA. Federal and state officials used this tool in 1999 to protect the Headwaters Forest in California, a significant habitat of privately owned old-growth redwoods (*The Economist*, 1999). The incentive to companies in such cases is that they then have a more secure economic environment, protected from ESA-inspired lawsuits launched by environmental groups. Negotiations concluded in 1997 in San Diego—a crucial region for national planning for endangered species—produced complex long-term arrangements for both protected land and development areas. According to Interior Secretary Bruce Babbitt, this particular agreement heralded a 'new era in American conservation. . . . Voluntary conservation partnerships on private land will be as important to America's natural heritage in our children's lifetimes as [the national parks and monuments and wildlife refuge system] in the early twentieth century was to us' (Ayres, 1997).

The network of agreements built in the US since the early 1980s has nonetheless been criticized by conservationists as inadequate to the task of protecting endangered species and habitats. Other critics have portrayed this record as an assault on constitutionally guaranteed rights. A basic problem for some environmentalist critics arises from compromises built into ESA practices. Sheldon (1998: 283, 300 ff.) argues that the Fish and Wildlife Service 'has developed some policies that trade species protection for landowner certainty.' In particular, ESA processes have applied disproportionately to small areas; they involve lengthy planning procedures, are restricted by lack of biological data, and create widened opportunities for stakeholder inputs and extended delays; uncertainty exists over the long-term enforceability of agreements; and adequate funding is lacking. Wilcove criticizes the underlying philosophy of the ESA as being 'purely punitive in nature'. There are fines and jail terms for offences related to species and habitats, 'but it provides no rewards or incentives to encourage good behaviour on the part of landowners' (1998: 277–8).

While these kinds of weaknesses are arguably remediable, the criterion of property rights leads others, including Canadian critics, to see the ESA as bad law. According to Smith and Smith (1997: 3–7, 23–6, 33), it 'places a disproportionate share of the burden on private landowners and causes owners to lose a substantial portion of the economic use of their land.' ESA listings and restrictions and its system of 'perverse incentives' deleteriously affect property values. They argue that landowners are denied the right to make changes to their property, including land that is purchased or inherited, and that the courts have tended to interpret the Act in ways that harm the interests of private landowners.

In several EU countries, as in the US, large numbers of NGOs are engaged in a variety of ownership and other land-management roles. For example, the Wiltshire Wildlife Trust in England, with around 10,000 members, manages about 2,000 acres on over 40 nature reserves in the county (WWT, 1999). Among national organizations, the Royal Society for the Protection of Birds, which claims around 880,000 members, manages some 120 reserves (*This Common Inheritance*, 1997: 39; Dwyer and Hodge, 1996; Evans, 1997: 208). Recovery programs in the UK have been looked on favourably by Canadian critics, who argue that US-style regulatory approaches are inappropriate for Canada.

These multi-stakeholder programs rest on several core features. First, the key is the voluntary principle. Agreements to protect a species or threatened habitat in Britain are arrived at through negotiations among stakeholders. Positive incentives are emphasized, to encourage good management practices in the interests of threatened species. Compulsory purchase of land by the government is a possible instrument after endangered species listing, but this is very rarely used. Second, the resulting frameworks and their implementation rest on a 'conservation triangle' of government, NGOs, and landowners. NGOs have key roles in data collection, overtures to landowners, and post-agreement management. Some municipal governments are also important players because of their significant land-use and zoning powers. Third, public monies are available to support programs, for example, in the form of matching funds or grants by the government to NGOs. Related programs connect these activities with others, such as the voluntary arrangements between government and farmers on 'environmentally sensitive areas', which in England account for about 15 per cent of agricultural land. Progress is often slow, however, and negotiations are time-consuming. Disputes are unavoidable. A recent case was conflict over the Syresham Marshy Meadows SSSI (site of special scientific interest) in Northamptonshire. A protection agreement was negotiated with the owner in 1987 and government funding was provided for wildlife protection, but a new owner in 1995 refused to accept the terms of the previous agreement (Young, 1998).

CONCLUSIONS

The fate of species at risk depends in part on the ways polities are organized. No model ecological polity, however, can provide guidelines for conservation in the real world. National societies are diverse. Lessons learned or models crafted in one jurisdiction are not necessarily transferable to others, even within the range of the comparable economic and political systems of Canada and other OECD countries. The brief exploration in this chapter nonetheless underlines two general points. First, problems of endangered species are serious throughout the Western world, amounting to crisis proportions in some regions. Second, these problems require a range of creative social and political approaches that rest on, but are not restricted to, ecologically sound government programs and legislation.

The experiences of other countries suggest both what is possible in the art of endangered species protection and also where some of the main difficulties lie. Of the countries discussed in this chapter, Australia's economic and political system has the most obvious similarities with Canadian structures. The Australian states, like the provinces, are fierce defenders of their roles in environmental and natural resource policy and resistant to federal government interference. As in Canada, questions of wildlife and habitat protection in Australia touch on many sensitive and complex issues of private landownership; Canberra's direct authority is restricted to a relatively small area of federal lands; and jurisdictional issues have increasingly had to accommodate Aboriginal land claims. Yet Australia has successfully crafted a national legislative strategy for species at risk underpinned by an active role for the federal government in collaboration with state governments. One factor for change has been a vibrant and articulate environmental movement at local, state, and national levels. Endangered species legislation also reflected critical political compromises in balancing the respective responsibilities of the states and the federal government, as well as scientific and political considerations in the listing and protection of species. Federal authorities in Canberra, varying with issues and circumstances and the ideological complexion of the government, have also been politically ready to make use of international environmental agreements to reinforce their domestic leadership role.

The comparative record shows that national legislation on species at risk is essential, despite problems of compliance and enforcement, as in the multiple jurisdictions of Australia and the EU, and the dangers of a political backlash, as was evident in the US in the early 1980s and again in the late 1990s. Good laws and related government programs also have indirect and longer-term consequences. In favourable circumstances, such laws and programs can serve a number of functions:

- stimulate conservation and species-recovery debates;
- provoke more sustained media interest;
- raise public awareness and inject important issues into elections;
- provide NGO campaigns and coalitions with focal issues;
- foster partnerships among governmental and non-governmental actors;
- bring in the courts in the clarification of important issues in key cases;
- facilitate government-sponsored data collection;
- strengthen the hand of endangered species agencies inside governments;
- generate attention to the need for expanded legislative and policy development in the future;
- reinforce the international environmental tasks of governments.

Yet, government programs depend on political will for their efficacy. They are vulnerable to continually changing economic and political conditions, including pressures on budgets and the volatility of public opinion. Governments, moreover, form only a part of the overall responses of societies to species-at-risk issues. Non-governmental organizations have become increasingly indispensable in Canada and other OECD countries. Apart from their traditional roles as environmental educators, they carry out crucial data-collection tasks, especially in relation to amphibians, invertebrates, and other relatively neglected groups of species. NGOs are also responsible for a significant share of the work of managing natural habitats, negotiating and developing constructive relationships with private landowners, and implementing action plans for species recovery.

Public discussion and media coverage in Canada of global biodiversity questions often tend to be directed towards problems of developing countries. While these are important, an assumption of this chapter has been that we also need more comparative attention to issues and developments in the grouping of Western nations of which Canada is part. As the federal government edges towards putting national species-at-risk legislation into place, its environmental credibility among these countries will grow. So, too, will Canada's capacity to play a more substantial international leadership role on questions of biodiversity.

NOTES

Research for this chapter was carried out in part with funding from the Social Sciences and Humanities Research Council of Canada, and I gratefully acknowledge this assistance.

1. In many countries outside of North America, the World Wildlife Fund is called the World Wide Fund for Nature.

2. In legislative plans announced in November 1999 the government said it intended to make mink and other fur farms illegal.

3. Critics argued nonetheless that resources made available for implementing the law were insufficient, and that the government failed to use key powers contained in the legislation (Bates, 1995: 307).

4. These areas are the external territories (including the Australian Antarctic Territory), Jervis Bay, federal national parks and protected areas, coastal waters (except areas under state and Northern Territory jurisdiction as a result of intergovernmental agreements), and the continental shelf and Australian fisheries zone (Bates, 1995: 304).

5. The most recent study was done by the United States Geological Survey (see USGS, 1999).

REFERENCES

Ayres, B.D. 1997. 'San Diego Council approves "model" nature habitat plan', *New York Times*, 20 Mar., A16.

Baker, Beth. 1999. 'Spending on the ESA—too much or not enough?', *BioScience* 49, 4: 279.

Baldock, David, et al. 1987. *The Organization of Nature Conservation in Selected European Countries*. London: Institute for European Environmental Policy.

Bates, G.M. 1995. *Environmental Law in Australia*, 4th edn. Sydney: Butterworths.

Bean, Michael J. 1983. *The Evolution of National Wildlife Law*, rev. edn. New York: Praeger.

Bennett, Graham, ed. 1994. *Concerning Europe's Natural Heritage: Towards a European Ecological Network*. London: Graham and Trotman.

Blenkinsop, Philip. 1999. 'The Netherlands is full of it—manure, that is', *Globe and Mail*, 28 Oct., C4.

Brown, Jacqueline Lesley. 1997. 'Preserving Species', *Journal of Environmental Law and Litigation* 12, 2.

Brown, G.M., and J.F. Shogren. 1998. 'Economics of the Endangered Species Act', *Journal of Economic Perspectives* 12, 3: 3–20.

Beebee, Trevor, and Jonathan Denton. 1995. *The Natterjack Toad Conservation Handbook*. Peterborough, UK: English Nature.

Chanwai, Kiri, and Benjamin Richardson. 1998. 'Re-working indigenous customary rights? The case of introduced species', *New Zealand Journal of Environmental Law* 2: 157–86.

Christiansen, Peter M., ed. 1996. *Governing the Environment: Politics, Policy and Organization in the Nordic Countries*. Copenhagen: Nordisk Ministerrad.

Colman, David. 1997. 'Next front: animals that aren't so cute', *New York Times*, 2 Feb., 41, 42.

Common Ground. 1999. Bonn: Federal Ministry for the Environment, Nature Conservation and Nuclear Safety.

Déjeant-Pons, M. 1998. 'La stratégie pan-européenne de la diversité biologique et paysagère', in M. Prieur and C. Lambrechts, eds, *Les hommes et l'environnement: quels droits le vingt-et-unième siècle?*. Paris: Editions Frison-Roche, 583–609.

Diana, Olivier. 1998. 'Natura 2000', *Naturopa* 87: 7.

Dobson, A.P., et al. 1997. 'Geographic distribution of endangered species in the US', *Science* 275 (24 Jan.): 550–3.

Dubasak, Marilyn. 1990. *Wilderness Protection: A Cross-Cultural Comparison of Canada and the United States*. New York: Garland.

Dwyer, Janet C., and Ian D. Hodge. 1996. *Countryside in Trust: Land Management by Conservation, Recreation and Amenity Organizations*. Chichester: Wiley.

The Economist. 1998. 'France's angry countrymen', 21 Feb., 51.

——. 1999. 'Saved, at a price', 6 Mar., 30.

Emneborg, Helene, and Frank Götmark. 2000. 'The role of threat to areas and initiative from actors for establishment of nature reserves in southern Sweden, 1926–1996', *Biodiversity and Conservation* 9, 6: 717–38.

Endangered Species Advisory Committee. 1992. *Australian National Strategy for the Conservation of Australian Species and Communities Threatened with Extinction*. Canberra: Australian National Parks and Wildlife Service.

European Environment Agency (EEA). 1998. *Europe's Environment: The Second Assessment*. Oxford: Elsevier Science.

Evans, David. 1997. *A History of Nature Conservation in Britain*, 2nd edn. London: Routledge.

Finkel, Elizabeth. 1998. 'Forest pact bypasses computer model', *Science* 282 (11 Dec.): 1968–9.

Forman, Richard T.T. 2000. 'Estimate of the area affected ecologically by the road system in the US', *Conservation Biology* 14, 1 (Feb.): 31–5.

Gaston, K.J., et al. 1998. 'Species-range size distributions in Britain', *Ecography* 21 (Aug.): 361–70.

Glaberson, William. 1999. 'Novel antipollution tool is being upset by courts', *New York Times*, 5 June, A1, A10.

Gottesmann, Jean. 1997. 'Nature conservation legislation in Switzerland', in Kiss and Shelton (1997: 211–19).

Griffiths, G.H., et al. 1999. 'Integrating species and habitat data for nature conservation in Great Britain: Data sources and methods', *Global Ecology and Biogeography* 8: 329–45.

Grossman, Margaret Rosso. 1997. 'Habitat and species conservation in the EU and the US', *Drake Law Review* 45: 19–49.

Halsbury. 1995. *Halsbury's Laws of Australia*, vol. 12. Sydney: Butterworths.

Hartmann, Joan Rae. 1981. The Symbolic Dimension of Politics: The Case of Endangered Species Legislation', Ph.D. thesis, Claremont Graduate School.

Kiss, Alexandre, and Dinah Shelton. 1997. *Manual of European Environmental Law*, 2nd edn. Cambridge: Cambridge University Press.

Line, Les. 1997. 'Navy moves to aid shrike', *New York Times*, 7 Jan., C4.

Littell, Richard. 1992. *Endangered and Other Protected Species: Federal Law and Regulation*. Washington: Bureau of National Affairs.

Loomis, John, and J. Chris Echohawk. 1999. 'Using GIS to identify under-represented ecosystems in the National Wilderness Preservation System in the USA', *Environmental Conservation* 26, 1 (Mar.): 53–8.

MacDonald, Janine. 1999. 'Senate set to OK green shake-up', *Age* (Melbourne), 23 June, 1.

Marlin, Richard. 1998. 'The external affairs power and environmental protection in Australia', *Federal Law Review* 24, 1: 1–28.

Morillo, Cosme, and César Gómez-Campo. 2000. 'Conservation in Spain, 1980–2000', *Biological Conservation* 95 (Sept.): 165–74.

Nadeau, Simon. 1999. 'The Australian approach: words of wisdom from abroad', *Recovery: An Endangered Species Newsletter* 13 (June): 5.

National Biological Service. 1995. *Endangered Ecosystems of the US: A Preliminary Assessment of Loss and Degradation.* Washington: Department of the Interior.

Porter, William F., and H. Brian Underwood. 1999. 'Of elephants and blind men: Deer management in the US national parks', *Ecological Applications* 9, 1: 3–9.

Roy, D.B., and T.H. Sparks. 2000. 'Phenology of British butterflies and climate change', *Global Change Biology* 6, 4 (Apr.): 407–16.

Sheldon, Karin P. 1998. 'Habitat conservation planning: Addressing the Achilles heel of the Endangered Species Act', *New York University Environmental Law Journal* 6, 2: 279–340.

Smith, Robert J., and M. Danielle Smith. 1997. 'Endangered species protection: Lessons Canada should learn from the US Endangered Species Act', *Property Rights Journal* 1, 1: 2–36.

This Common Inheritance. 1992, Cm 2068; 1994, Cm 2549; 1996, Cm 3188; 1997, Cm 3556. London: HMSO.

United States Geological Survey (USGS). 1999. *Status and Trends of the Nation's Biological Resources.* Washington: US Government Printing Office.

Van der Zande, A.N., and A.R. Wolters. 1997. 'Nature conservation in the Netherlands', in Kiss and Shelton (1997: 219–28).

van Heijnsbergen, P. 1997. *International Legal Protection of Wild Fauna and Flora.* Amsterdam: IOS Press.

Wilcove, David S. 1998. 'The promise and the disappointment of the ESA', *New York University Environmental Law Journal* 6, 2: 275–8.

Wiltshire Wildlife Trust (WWT). 1999. www.wiltshire-web.co.uk

Young, Baroness. 1998. Letter in *Country Life*, 30 July, 5.

International Initiatives, Commitments, and Disappointments: Canada, CITES, and the CBD

PHILIPPE LE PRESTRE AND PETER STOETT

INTRODUCTION

The need to mitigate the rapid loss of biodiversity on a global scale provides states with great incentive for co-operation. There is no doubt that we are currently living in an age of human-induced mass extinction, with over 1,000 species lost every year (Tuxill and Bright, 1998; Baillie and Groombridge, 1996). Nor is this a new issue; we have been aware of how urgent a problem this is for over two decades (Fenton, 1995; Eldredge, 1991; Ehrlich and Ehrlich, 1982), and it is widely accepted that loss of biodiversity will persevere as one of the most important policy challenges facing governments in the near future, both domestically and internationally (McNeely, 1997). Though this book focuses largely on species found within Canada, it is self-evident that Canadian responsibilities extend beyond its borders and that the conservation of wildlife on as global a scale as possible remains imperative.

Canada maintains many international commitments for the protection of global wildlife. This chapter will examine two of them in detail, the Convention on International Trade in Endangered Species of Wild Fauna and Flora (CITES) and the Convention on Biological Diversity (CBD). Both have received oscillating levels of media attention over the past decade, but Canada has played a firm role in the evolution of one and was instrumental in the establishment of the other. While CITES is a much more specific treaty aimed at curtailing threats to wildlife from trade only, the CBD goes far beyond this and has a much more diffuse role regarding the preservation of habitat. We will discuss CITES first, because it preceded the CBD and concerns a narrower range of issues, before moving into an examination of Canada and the CBD. Several themes link the two, however: both are fragile institutions, susceptible to challenges to their legitimacy; both must integrate often wildly divergent ideological perspectives to function; and both require not only international commitments but—and herein lies the principal complication, especially in federal states such as Canada—accompanying domestic legislation as well.

The fifth goal of the Canadian Biodiversity Strategy is to 'work with other countries to conserve biodiversity, use biological resources in a sustainable manner and share equitably the benefits that arise from the utilization of genetic resources.' While the Canadian commitment to CITES and the CBD can be viewed as the two pillars of this effort, there are several other agreements of note. Those dealing with flora include the International Plant Protection Convention, the International Convention for the Protection of New Varieties of Plants, the North American Agreement for Plant Protection, and the International Tropical Timber Agreement (ITTA). And there are even more related to fauna, ranging from the Convention on the Conservation of Antarctic Marine Living Resources to the Agreement on the Conservation of Polar Bears and the Agreement between Canada and the United States on the Conservation of the Porcupine Caribou Herd. Canada is also a ratified signatory to the important Convention on Wetlands of International Importance, Especially as Waterfowl Habitat, commonly known as the Ramsar Convention.

In terms of related international organizations, the International Union for the Conservation of Nature (IUCN, or World Conservation Union) remains an important forum for international co-operation; its Species Survival Commission produces the Red List, which catalogues the world's most endangered species and is widely considered an authoritative source. In addition, Canadian zoos are involved in extensive efforts, based on international co-operative management, to save highly endangered species ex situ (see Wiese and Hutchins, 1994). Other international arrangements, such as the Global Environment Facility, have a specific mandate to reduce biodiversity loss (Fairman, 1996); and the United Nations Environment Program (UNEP) is involved in a number of areas, including a partnership with the IUCN. Another bilateral/multilateral instrument involving both governments and non-governmental organizations is 'debt-for-forest' swaps (Jakobeit, 1996), though Canada has had limited involvement in these so far. Other mechanisms, such as agreements between northern pharmaceutical companies and southern governments to conserve rain forests for the purpose of 'bioprospecting' (Mulligan and Stoett, 2000), are taking shape, but these are driven largely by American, European, and Japanese concerns.

But no treaties or conventions involving Canadian input cut across species and link habitat protection issue-areas as explicitly as CITES and the CBD, respectively. This makes them of interest not only to those concerned with national wildlife conservation, but to scholars of international relations as well. We are beginning to see the emergence of literature addressing the question of whether regime participation has a discernible impact on domestic policy formation (Schreurs and Economy, 1997), a question of particular relevance in the age of globalization. A 'reflectivist' approach argues that states are 'identity-seeking actors' and that 'part of what states do in defining themselves is to enter into lasting relationships with other states and to form institutions' (Weber, 1997: 234). As we examine the internationalization of public policy, we ask what institutional affiliations impact on domestic governance. This approach is increasingly popular among scholars of economic policy and other areas (see Johnson and Stritch, 1997, for partial examples). Has the interactive diplomatic context of international regime networking helped shape a unique identity for individual states? Does it really encourage policy convergence at the domestic level? And has state participation modified the outcome of the regime itself: its structure, its inner controversies and prevailing norms, and its legitimacy?

In this light, CITES and the CBD share a fundamental characteristic: they demand both external and internal policy adjustments and initiatives from their adherents. There is specific corresponding enabling legislation necessary to implement the agreements at the domestic level, including, in the case of CITES, active law enforcement. And there are controversial decisions to be taken abroad at the international meetings of the conventions, which will over time come to shape the image of Canadian environmental foreign policy. Leaving either institution at this time (as, for example, Canada left the International Whaling Commission in 1982) would be unthinkable.

So the question is whether Canada's actions, at home and abroad, reflect a commitment to the development and maintenance of the regime in question. In both cases the results are mixed, but the lack of domestic progress on species preservation has ultimately proven disappointing in the context of Canada's broader international commitments. Further, there is a soft irony at work here in the distinctions between the formation and implementation stages of the international conventions. While Canadian involvement in the formation of CITES was limited, it has certainly increased over the years and, arguably, has resulted in concrete changes in national legislation. Canadian involvement in the formation of the CBD was unusually strong, but this has faded with recent developments. This is probably because, of the two, only the CBD has much broader implications for industry, property rights, and intellectual property rights.

We will proceed with a rudimentary description of CITES. Next we discuss the domestic dimension of its implementation, which is absolutely fundamental to the success of the regime. If states do not monitor their borders for trade in endangered species then highly lucrative poaching becomes virtually risk-free. Next, we explore the international dimension of the regime: the ideological divide that has grown between animal rights activists and conservationists; questions of funding commitments; and the larger political/economic context of free trade. We close our discussion on CITES with reference to ongoing issues affecting both the regime and Canadian foreign policy. Primary among these is the need to move beyond CITES—to protect not just individual species from poaching, but their habitat from anthropogenic incursion. This takes us directly to a lengthier discussion of the CBD.

While CITES was an American initiative in the early 1970s (the Nixon administration has a surprising environmental record), the CBD was at various stages opposed by the United States and countries such as Canada filled the gap. Canada even provided additional funding to establish the home of the secretariat in Montreal. Thus, our section on the CBD will deal with Canadian involvement at the formative stage of the regime in much more detail. As importantly, the CBD raises the thorny issue of sharing access to and the benefits of biodiversity, whereas CITES does not go beyond a largely regulatory approach. We will see also that Canada's initial support for the

CBD has waned somewhat with time, and the need to ensure Canadian vigilance on international wildlife conservation issues is an ongoing one.

CITES

Wildlife has been traded throughout history.[1] The products have provided subsistence for many trappers and traders; buyers generally apply them as status symbols, aphrodisiacs, fashion accessories, clothing, and, in their living state, pets. This century, however, saw the meteoric rise of such trade, to the point where it threatens the very survival of many exotic species worldwide. Previously, cost prohibited large-scale consumption of such rare commodities, but after World War II the 'massive worldwide growth in the transportation and communication infrastructures that facilitate trade [and] the affluence of the developed nations combined with global population growth' have led to the present situation (Slocombe, 1989: 20). Coupled with habitat destruction, hunting and poaching can be fatal for species on the edge of extinction. Illicit trade in species remains one of the most pernicious causes of wildlife decline, however, with an estimated value between $10–20 billion (US) a year (Webster, 1997; Burgess, 1992; Fitzgerald, 1989).

Previous attempts to control the international utilization of wildlife were quite explicit in their goal: to preserve the industry in question, not the mammals themselves. In Canada's name, Great Britain signed the Fur Seal Convention of 1911 (joined by Japan, Russia, and the United States), which banned all pelagic commercial seal hunting in the North Pacific (see Dorsey, 1991). A Convention for the Regulation of Whaling was signed in Geneva in 1931 (later replaced by a more robust treaty but nonetheless one designed primarily to maintain the whaling industry). In November 1933, the Convention Relative to the Preservation of Fauna and Flora in Their Natural State was signed in London. It was by all accounts a weak convention, designed largely to conserve the African safari experience. But it provided the basis for subsequent international efforts, as well as a formal attempt at classifying the most endangered species in two categories, Class A (including gorillas, Madagascar lemurs, white rhinoceros, and young elephants) and Class B (including chimpanzees, giraffes, wild ostrich, and older elephants). Class A listing provides the most stringent protection.

It was quite some time before trade in species and parts became recognized as a fundamental threat to their existence. Under a largely American initiative, CITES was adopted in 1973 in Washington, DC, and entered into force in 1975. There are currently over 140 party states to this Convention. Its main forum for decision-making is a Conference of the Parties (COP), which meets once every two to three years. These meetings are high-profile events, generating widespread media coverage and the attention of hundreds of non-governmental organizations (NGOs). However, media coverage is usually limited to the debates on the more popular species, and the real work of CITES takes place on many other levels, including that of national governments, regional organizations, and global conservation networks. In addition to elephants and whales, for example, some 15 species of tropical timber (see Amilien, 1996) are listed in CITES, as are hundreds of endangered flowers and birds. Much as the Endangered Species Act in the US can serve to conserve habitat by protecting individual species, CITES has the potential to encourage governments to adopt habitat preservation areas in return for trading allowances and/or the internal justification of anti-poaching legislation. It represents a unique political space where governments, NGOs, scientists, industry lobbyists, and others converge and diverge as they assert their respective agendas.

Out of this has evolved an ideological split between 'preservationists' and 'conservationists' (often known as *utilizationists*). The former would eradicate most forms of harvesting endangered wildlife altogether, arguing it is time to move into a new environmental awareness paradigm. The latter insist that species should be protected only to a point and that human consumption (for eating, decoration, or other forms of use) should be allowed to continue (see Stoett, 1997). In policy terms, the debate revolves around the appropriate 'listing' of species: within the CITES regime are three categories of placement, each with different consequences. Appendix I is for species threatened with extinction; with few exceptions, most trade in products derived from those listed is prohibited. Appendix II applies to species 'which although not necessarily now threatened by extinction may become so unless trade in specimens of such species is subject to strict regulation in order to avoid utilization incompatible with their survival'; an export permit from the exporting country, implying a finding of 'non-detriment' by a national scientific authority, is necessary for trade in these species. Finally, Appendix III listings result from a country's unilaterally declaring species on the list and providing export permits for exports. Determining the appropriate category of various endangered species has become an issue of surprising political importance.

However, all of these regulations are inconsequential unless there is an effective compliance mechanism in place for the regime itself. Import and export restrictions are often violated; not all signatories are in clear compliance; and the regime remains hostage to the threat of withdrawal routinely made by several African and Asian states that dislike current conditions. A further question relates to the reciprocal effect of participation in international organizations: does such participation affect the domestic policy of involved states? To date, no academic work looks at the long-term impact of Canada's participation in the CITES regime, and even general work on the cross-national impact of CITES is lacking or outdated (Emonsds, 1981). Canadian legislation, however, has clearly reflected the conditions imposed by a commitment to CITES.

The Domestic Dimension

Many popular and many more obscure species found in Canada are listed in the CITES appendices. The polar bear, river otter, and burrowing owl are examples of Canadian species on Appendix II, as are the lynx, bobcat, and cougar. The black bear is an example of a Canadian species listed for look-alike reasons, a controversial category that requires export permits (the majority of which have been obtained by

American hunters returning trophies to the US; recent regulatory proposals would eliminate the need for these permits, as well as those for other household effects). The walrus is on Appendix III.

There is no doubt that CITES stands as an example of how domestic legislation reflects official adherence to an international treaty. Generally, CITES had been implemented under the General Import Export Act administered by Foreign Affairs. In December 1992 the Wild Animal and Plant Protection and Regulation of International and Interprovincial Trade Act replaced the Game Export Act. The Act was brought into force on 14 May 1996. It is based in part on CITES, which Canada ratified on 10 April 1975. The Act is administered by the Enforcement Branch of Environment Canada and carried out by five regional offices (Pacific and Yukon, Prairie and Northern, Ontario, Quebec, and Atlantic). Subsection 10(1) gives the Minister of the Environment the ability to 'issue a permit authorizing the importation, exportation or interprovincial transportation of an animal or plant, or any part or derivative of an animal or plant'. The powers afforded are in fact quite sweeping. Subsection 13 says the suspected item can be 'detained by an officer' if it is transported out of or into another province or another country. Subsection 14(1) says that an 'officer may at any reasonable time enter and inspect any place in which the officer believes, on reasonable grounds, there is any thing to which this Act applies'; and this includes the right of seizure, though a warrant is needed for searching a dwelling place. The maximum fine for violation is $300,000; the maximum prison sentence is five years.

Implementation duties are diffuse. Generally, Parks Canada monitors wildlife trafficking in national parks. Provincial natural resources ministries monitor wildlife in their provincial parks and issue hunting and export permits (with the exception of Alberta, which relies on the federal government). Both federal and provincial agencies conduct spot checks or routine inspections of wildlife businesses, conduct investigations, and gather intelligence on possible illegal activities. The Royal Canadian Mounted Police and even Interpol are also involved. However, there is little overall coordination of activity, and resources suffer from chronic underfunding.

In principle, CITES administration is centred in the Canadian Wildlife Service of Environment Canada, with head offices in Hull, Quebec. Though the CWS plays an important role, particularly in attending the COPs and deciding on Canadian voting patterns, many would argue that other agencies are more involved in implementation at home. This would include the many federal departments and agencies that deal with CITES issues, such as Fisheries and Oceans Canada, Agriculture Canada, External Affairs and International Trade, the RCMP, Canada Customs and Revenue Agency, and the Canadian Food Inspection Agency. The Canadian Museum of Nature also provides an advisory role.

Given the importance of hunting and trapping in Canadian history and Aboriginal life it should come as no surprise that a wide variety of people are affected by wildlife trade controls. Canadians involved in exporting wildlife include trappers, outfitters, hunting guides, and taxidermists, and large greenhouse operations also require permits for exports of plants such as cyclamen and cacti. In 1992, there were 10,396 export permits issued: 5,742 for black bear and 4,654 for other species covered by CITES (Appendix II). The vast majority of black bear export permits were from Ontario (BC was a distant second); most of the 'other' species came from Quebec (Outspan Group, 1993: 12). Most of the black bear (*Ursus americanus*) permit holders were American hunters seeking to return with the carcass, though the main concern has been poaching for gall bladders for export to Asia. Eliminating the need for black bear permits would greatly reduce the administrative burden of CITES: 'For hunting trophies alone, some 8,600 black bear and 100 sandhill crane CITES export permits would no longer be needed by U.S. hunters taking their legally harvested animals back home.' The report adds that the current spring hunting closure in Ontario should not affect overall numbers since 'hunters will shift their efforts to the fall season' (Environment Canada, 1999: 7).

Though it is impossible to estimate the number of illegal imports and exports of endangered species, it is apparent that an effort is made to apprehend such shipments. In 1996–7 Environment Canada completed 4,141 inspections, 209 investigations, 12 prosecutions, and 4 convictions. This would suggest some progress has been made since an important study conducted by Douglas Hykle in 1988 and an article in the *Globe and Mail* (20 Dec. 1994, A4), which claimed Canada was being used extensively to 'launder' species ('import them illegally and then re-export them with Canadian papers so they can be easily sold in other markets'). But there is no doubt that the CWS remains dreadfully underfunded and that further co-ordination among the many law enforcement agents involved in CITES implementation is necessary. However, this is a complicated task, demanding better (and more expensive) training of customs agents and the continued legitimacy of the CITES regime itself. Modern technology, such as forensic testing to obtain the species identity of suspicious shipments, is still beyond the financial reserves of states such as Canada and the United Kingdom, let alone the exporting countries in the southern hemisphere. There is some hope that the use of coded-microchip implants for marking live animals in trade might help tracking services, but for small wildlife parts (many of which have already been converted to household goods or decorations) this would not be feasible.

The International Dimension
Saving endangered wildlife outside of Canada has been the main focus of CITES activity at the COPs. Several high-profile NGOs have used whales and panda bears as symbols of their efforts, and the often celebrated campaign to ban international trade in ivory has been a constant CITES issue. The ban was implemented in 1989, though it has been partially lifted to allow some southern African states to sell stockpiled ivory to Japan. This was a highly controversial development, reflecting the high status accorded to elephants by the NGO community, on the one hand, and the legitimate claims to developmental rewards for conservation practices, on the other (see Stoett, 1997). The most recent COP continued to reflect divisions on the theme of wildlife use,

as whaling states were defeated in their efforts to delist several species of whales and diverging states agreed basically to maintain the status quo on elephants.

The appearance of a species on Appendix I represents two things at once: a recognition of a dreadful threat to the very survival of the species and a political victory for those countries and NGOs that have lobbied hard to get the necessary two-thirds majority vote. It is at once a scientific appraisal and the conclusion of intensive, often transnational, lobbying. Though there is no comprehensive record of Canadian voting patterns at CITES, it is safe to say that Canada, with its predominant conservationist/utilizationist stance, has been a cautious voter in CITES procedures. This means delegates are reluctant to upgrade species without solid evidence of their relevance to the Convention and their endangerment; but it also implies that delisting is not often accepted. Canada has acted to protect domestic economic and/or cultural interests at every turn. In 1985, Canada vigorously opposed the prospect of listing hooded seals in Appendix II, 'on the grounds of inadequate biological justification and their impact on native and non-native sealers' (Slocombe, 1989: 26), especially after the EU's banning of seal products in 1982. Delegates to CITES conferences cannot expect to satisfy all those who lobby them, since there are such wide gulfs between conservationists and preservationists.

Canada does meet its funding obligation to the CITES administration. Before 1987, CITES was funded largely through the UNEP, but after that the parties' funding was allocated according to the UN system linking GNP with contribution. For example, using US dollars, Canada's allocation was 3.11 per cent of the total CITES budget, which for the 1996–7 period was $143,814 annually; for the 1998–2000 period, Canada's annual payment was $139,898. France, at 6 per cent, contributed $277,455 annually; Japan, at 12.45 per cent, $575,719; and the United States, at 25 per cent, $1,156,062. These are very small amounts given that global illegal trade may be worth well over $10 billion globally. Most states are evaluated at 0.01 per cent, which is less than $500 a year. In fairness, however, these states may have a hard enough time incurring the extra expense of sending representatives to the biennial meetings. The CITES overall budget remains very modest, the main expenditures going towards professionals hired for commissioned studies, staffing, subcontracts, and travel. The 1999 budget was less than $5.2 million. This represents a fraction of the CITES regime costs, since governments are expected to enforce trade regulations at the national level. Again, this is a difficult proposition for states with limited resources to invest in this activity. Anti-poaching efforts in Kenya and elsewhere have attracted international funding and attention, but the day-to-day routine of patrolling borders and customs checking is still often shrouded in mystery. Officials may be susceptible to bribes or even engaged in the trade itself. A long line of value-addition characterizes the wildlife trade, with severe markups on products that find their way into northern and Asian markets.

CITES faces the problem of stemming demand for what are, in many cases, still highly desirable objects such as gorilla parts, gall bladders, and rhinoceros horn. In many locations the preservationist agenda, which some say CITES has moved closer to adopting, is viewed as outright cultural imperialism. (It is largely perceived this way in Canada's North as well.) The hope that banning trade in certain species will ultimately lead to reduced demand is countered with the development of a black market for the product and the slowness with which local customs change (and there is no clear consensus that trade bans reduce mortality). Poaching remains lucrative for the middlemen who orchestrate the trade and make a calculated decision weighing potential costs of detection against the profit of success.

Along with attending the COPs, contributing expertise to CITES-related species inventories, and helping to publicize the regime, Canada has demonstrated a long-term commitment to CITES, regardless of potential problems arising from conflicts between preservationists and conservationists. There can be little doubt, given the resource-use history and current dispositions in Ottawa, that Canada has generally fallen within the latter camp, despite the presence of several vocal NGOs within Canada promoting its opposite. The strain is noticeable beyond the CITES biennial meeting level. In particular, changing attitudes towards the fur trade in Europe and elsewhere have had a strong impact on the industry for both Native and non-Native Canadians. Since January 1995, European Union countries have forbidden importing animals or animal parts listed as endangered species unless the European Commission is assured that the country of origin prohibits leghold traps and that the trapping methods used meet certain standards. The fur industry was upset that it had to obtain permits for lynx, wolf, and otter fur because they are look-alikes of endangered species. However, Canadian officials have given negotiations with the EU some priority and, along with the International Fur Trade Federation, 'appear to have been successful in preventing CITES from being merged with European concerns and thereby becoming a potential barrier to legitimate trade' (Outspan Group, 1993: 114–15).

There are tensions also between the CITES regime and the broader neo-liberal trade regime carved out of the General Agreement on Tariffs and Trade and the World Trade Organization (WTO). On the regional level, the North American Free Trade Agreement (NAFTA) could become a major issue as well. Some states have claimed that CITES can be used as a hidden trade barrier. For example, the Japanese required a CITES permit for Canadian exports of artificially propagated vanilla beans in an effort to support their own industry. (Though not a CITES-related decision, the infamous tuna-dolphin case has led environmentalists to doubt the ability of the WTO to make decisions that take the environment into consideration.) This is why it is important, and complicated, to have a strong CITES regime with legitimacy abroad: once a species is on the appendices there is no need for immediate political decisions or for the WTO to rule on trade admissibility. The critical nature of biodiversity reduction seems to have convinced most governments that some things justify violating the push towards freeing global trade; however, trappers and hunters often feel they are subjected to impositions from a far-off Western middle-class environmentalist agenda in the process.

CITES represents a fascinating political space where countries are forced to reconcile their own national interests, in particular those of their wildlife exporting industries, with a genuine concern for the mass extinction currently taking place across the planet. In this sense it is not unlike many other international agreements, including arms control provisions. There is a visible need for verification, since signatories that do not enforce the agreement can actually profit at the expense of those that do. And members have to adjust, often suddenly, to new circumstances such as the dissolution of the Soviet Union (which triggered a new era of poaching in the area). There is the additional problem of non-members, such as Saudi Arabia, Turkey, and South and North Korea, forming trade bridges.

Ongoing Issues

CITES remains threatened with a philosophical split between preservationists, who would use the treaty as a way to stop all trade in certain wildlife parts, and conservationists, who recognize the need to conserve species but do not have ethical qualms about utilizing nature. One can argue that both sides of this debate share a common concern for habitat protection, and that is where international efforts must be focused in the future. In the meantime, however, the pressures resulting from continued demand for wildlife parts necessitate a regime designed to limit trade in endangered species, and here the philosophical division is much harder to bridge.

For its part, Canada must continue to display a commitment to the CITES regime at home and abroad. This can be done by continuing to improve the implementation process while taking care not to alienate those in the affected industries. As for international verification procedures, Canada must continue to support the Wildlife Trade Monitoring Unit, part of the Species Unit within the World Conservation Monitoring Centre in Cambridge. The Species Survival Commission of the IUCN, which helps to co-ordinate many initiatives to preserve endangered species worldwide, is in need of continued assistance as well (this is rather complicated because of the variety of national approaches to zoo funding). Along with the World Wildlife Fund (WWF), the IUCN also supports the Trade Specialist Group TRAFFIC Network, founded in 1976 to monitor trade in wild plants and animals and to report to CITES. The acronym TRAFFIC is self-explanatory: Trade Records Analysis of Fauna and Flora in Commerce. Put bluntly, if exports do not equal imports, something is wrong.

Beyond this, several intergovernmental connections are maintained to deal with wildlife trade regulation, including links with the US Fish and Wildlife Service, the Mexican federal Attorney General's Office for Environmental Protection, the World Customs Organization, and Interpol. To date, Canada has not signed the Bonn Convention on Conservation of Migratory Species of Wild Animals, which uses the CITES appendices to define 'endangered'; it would be a positive step to do so.

The eleventh CITES Conference of the Parties was held in April 2000 in Gigiri (near Nairobi) at the UNEP facilities there. The event captured limited media attention, though the raucous debates over proposals to delist the African elephant (which were not passed) did generate some coverage. NGOs were particularly visible at the meeting, since high-profile species such as panda bears and whales can draw public attention. One cannot overestimate the NGOs' 'tactical use of the media to maintain pressure on media-conscious politicians' (Doern, 1993: 11). At the same time, however, this must be measured against the weight of Aboriginal groups, which have come to perceive distant regulatory organizations such as CITES and, in particular, the International Whaling Commission as intrusive and often misinformed. There was a large schism between those who valued the Appendix I listing of certain whale species and elephants as a symbolic victory and those who saw it as punishing users of wildlife. Other high-profile species under discussion included the hawksbill turtle and the great white shark. Again, Canada's voting record was cautious, and many of the delegates expressed concern with the difficulty of making informed decisions with the imperfect information available.

Indeed, the wildlife issue cannot be viewed aside from the need to integrate local development and resource management plans with a global approach. As such, it is inherently a human rights issue affecting the future of indigenous people around the world (Colchester, 1994; Slocombe, 1989). This gives rise to critical examinations of what the CITES regime can accomplish if it is driven by a northern agenda, reflecting a broader critique of global environmental management (see Sachs, 1993). This will be of importance to Canadians as we struggle to deal with policies regarding the hunting and trapping activities of the Inuit and other indigenous peoples. There is room within a complex network such as CITES for the encouragement of local lifestyles with sustainable approaches to wildlife use. Hopefully, Canada can lead by example by improving the domestic dimension and sustaining the international dimension of CITES implementation.

CITES does not have direct responsibility for habitat protection, but the policy-issue network involved in CITES debates and developments must do all it can to thrust the habitat question to the front and centre. There are severe limitations, however, to the level of CITES effectiveness on the habitat question, since most habitat destruction does not involve trade of species or parts across borders (for example, forestry and fisheries are the two industrial sectors affecting CITES species). In this light, it becomes obvious that the linkages between CITES and several other international conventions must be fostered and strengthened; in particular, CITES staff deal with questions related to the WTO, the ITTA, NAFTA, and, of course, the Convention on Biological Diversity.

THE CONVENTION ON BIOLOGICAL DIVERSITY

The role Canada has played in relation to the Convention on Biological Diversity illustrates a mixture of environmental vision and domestic constraints followed by an apparent inability to capitalize effectively on past leadership and by slightly two-minded policy initiatives.[2] On the one hand, the Canadian contribution to the negotiation of the CBD was in many ways remarkable. On the other hand, Canada has lagged behind in implementing the accord and has since abandoned its role as leader—except in one or two issue-areas—to reach a nadir in 1999 when, prodded by the United States, it took the lead of the 'Miami Group' in opposing a biosafety protocol that would put unacceptable constraints on the activities of its seed growers. From being a leader, Canada became a laggard and threatens to be a blocker.

This change of attitude in itself would deserve scrutiny. Numerous hypotheses could be advanced to explain it, from the personalities at the top of the Department of the Environment and the change of Prime Minister, to the fusion of foreign affairs and commerce into a single department, the constraints of a federal system, the strength of domestic interests, and the absence of an overriding consensus around a core set of propositions and solutions to biodiversity-related issues. This section, however, covers a happier story, namely the contribution that Canada has made to the negotiation of the CBD, from 1987 when UNEP initiated the process to 1994 when the second Conference of the Parties selected Montreal as the seat of the secretariat. A few final words will bring the story forward to 1999.

The Origins of the Biodiversity Convention

The Convention on Biological Diversity, signed in 1992 at the Rio Summit, is the legal answer to a variety of concerns, among which were the fragmentation of international conservation-oriented agreements that reflected different norms, objectives, means, and ends; the urgency of preventing the extinction of thousands of plant and animal species, known and unknown; and the need to respond to the emergence of biotechnologies offering new medical promises as well as new sources of revenues. Thus, during the 1980s, different groups began advocating in favour of a convention that would answer their particular concerns, from the creation of more parks to the protection of farmers' rights, from access to genetic resources to sustainable development (Swanson, 1999).

The Legal Community

Access to and exploitation of biological resources have been codified domestically for centuries. Externally, international efforts to protect animals and plants began 100 years ago, with the Convention for the Preservation of Wild Animals, Birds, and Fish in Africa signed in London in 1900 (McCormick, 1995). Since then, dozens of international treaties regulating specific aspects of conservation have been adopted. They reflected the dominant scientific, economic, political, and normative concerns and knowledge of the time, resulting in a hodge-podge of agreements that embody various goals and philosophies of conservation. Some focus on a particular species usually of economic value, others on an ecosystem (e.g., wetlands), still others on plants and animals in general. Some regulate a practice (e.g., driftnet-fishing), others many species sharing similar behaviour (Bonn Convention on Migratory Species; highly migratory fish). Some seek to protect a species (blue whales), others to manage its sustainable exploitation, and another, as we have just seen, trade. Confronted with this logjam, legal scholars, in the early 1980s, argued in favour of consolidating and harmonizing current texts, taking advantage of the emerging principles of environmental conservation and sustainable development. Their goal was to build a framework that would govern the interpretation, implementation, and evolution of existing and future agreements on species conservation.

Scientists and Environmentalists

At the same time, problems linked to the loss of biodiversity also benefited from renewed attention from the media, scientists, and environmentalists. Historically, scientists and environmentalists have often been at odds, but here they joined forces in the wake of several public controversies, among which the destruction of tropical forests clearly stands out. This issue gave rise to large political coalitions composed of biologists concerned with species loss, ecologists worried about habitat destruction, geo-chemists wondering about the impacts of disruptions of geo-chemical cycles, ethnologists deploring the disappearance or acculturation of indigenous populations, social activists eager to improve the fate of poor peasants and unequal property rights and land tenure, and media happy to tell stories of greed, destruction, and despair.

The World Conservation Union (IUCN), which already was promoting a single legal instrument to protect the world natural heritage, and the World Wildlife Fund, as well as rock stars, other public personalities, reporters, and scientists, pressured political leaders to pay attention (see, e.g., O'Connor, 1998). Scientists became increasingly aware of their ignorance of the richness of the very ecosystems that were being transformed. At the same time, they had to revise their earlier ideas about conservation, preservation, and ecosystem stability. In particular, their thinking about conservation paid more attention to the need to integrate the human variable into conservation policy. The resulting discourse was dramatically different. The rationale for species conservation was no longer based only on what was known but also on what was unknown. Protection applied not only to valuable or cute species, not only to known species in general, but to unknown ones as well. Ecosystems have to be protected precisely because we do not know—even cannot know—their worth. Scientific ('unknown species richness'), utilitarian ('unknown medicines'), and humanistic ('rich indigenous cultures') arguments prevailed.

Although the concept of ecological diversity had been around for some time, the conservation of individual (primarily animal) species mobilized public attention until the late 1980s. Deeply concerned both with the rate of species extinction and the destruction of habitats, biologists started using the term 'biodiversity' from 1986 onward. The significance of this change is not to be underestimated. The scope of the problem became all-encompassing, referring to the full variety of life on earth and pointing to a more holistic approach that included species, genes, ecosystems, and cultures. The 1992 Convention on Biological Diversity symbolizes this passage from the panda to the abstract concept, which entailed not only the adoption of an ecosystem approach but also taking into account whole patterns of interactions between humans and nature.

Industry

The concerns of the pharmaceutical industry stemmed from a renewed interest in the potential benefits that could be derived from the exploitation of biodiversity as biotechniques improved. Supported by several industrialized states, industry, including seed growers, wanted to maintain access to ex situ and in situ gene pools. In their mind this access had to be free although not necessarily free of charge or devoid of conditions. In addition, industry wished to protect the intellectual property rights of plant breeders over new varieties they may have developed (as recognized by the 1961 Convention of the International Union for the Protection of New Varieties of Plants) while controlling any extension of the rights accorded farmers by the 1986 International Undertaking on Plant Genetic Resources for their role in the conservation of genetic resources used in the seed industry.

These diverse strands were helped along by UNEP, which sought to lead an international movement behind a new convention. Its executive director, Mostafa Tolba, had been very active in the 1980s in facilitating the emergence of new environmental concerns on the international agenda and the negotiations of subsequent legal instruments. After enabling regional seas agreements, UNEP had played a significant role in bringing about the Vienna Convention on the Ozone Layer (1985) and its Montreal Protocol (1987). In the late 1980s it was actively engaged in devising new instruments to address the trade in hazardous waste and control of greenhouse gases.

The early proposals for a convention came from the IUCN. Its 1980 World Conservation Strategy, launched in collaboration with the UN's Food and Agriculture Organization, UNESCO, and the WWF, sought to pursue sustainable development through the conservation of biological resources and provided the foundation for future discussions and efforts to reconcile protection and development (IUCN, 1980). In contrast to the declaratory 1982 World Charter for Nature—which mentioned the need to protect biological *diversity*—the Strategy identified specific national and international measures that would stop the destruction of natural resources (Caldwell, 1996). The 1987 report of the World Commission on Environment and Development, *Our Common Future* (the Brundtland Report), also stressed the need to use natural resources sustainably and called for a biodiversity convention. At Mostafa Tolba's suggestion, the UNEP Governing Council, in 1987, decided to establish a working group of technical and legal experts to examine the relevance of IUCN ideas for the negotiation of an international legal instrument to protect biodiversity. From late 1988 on, an ad hoc expert group met regularly to write up a draft of what would eventually become, four years later, the CBD after a short period of negotiation by an Intergovernmental Negotiating Committee. One hundred fifty-six states formally signed the Convention at Rio (Raustiala and Victor, 1997), which entered into force 18 months later, on 29 December 1993, after ratification by 30 countries. Canada was the fifth to sign and the sixth to ratify the Convention. At the end of 2000, there were 178 parties to the Convention.

The character of the CBD was to be shaped by the different concerns of the various constituencies described above, as well as by the very dynamics of the negotiations and the evolving context of international environmental co-operation. In particular, the Group of 77 (G-77) pursued the traditional objectives of the now defunct new international economic order (NIEO) diplomacy of the 1970s and early 1980s, such as protection of national sovereignty, transfer of financial and technical resources, capacity-building, participation in international decision-making bodies, market access, etc. The result went far beyond the early intentions of the IUCN and of those who were concerned first with protecting species or ecosystems. From an initial focus on conservation, the Convention rapidly became a different instrument geared towards sustainable development. Without that enlarged scope, there would not have been any agreement; but that very scope also created tensions among the different goals of the regime. As in other instances, such as the Rio Declaration, the result was not so much a compromise as a juxtaposition of different priorities.

The CBD has three basic objectives:

(i) the conservation of biological diversity, including all plant and animal species as well as their genes and the ecosystems to which they belong—in situ conservation is favoured, notably through the creation of national parks and protected areas;

(ii) the sustainable use of the components of biodiversity through programs aiming at both economic development and the protection of the biological resources on which it is based;

(iii) the fair and equitable sharing of the benefits derived from the use of genetic resources. This goal invites developed countries to devise ways of compensating the biologically rich developing countries for the commercial use of their genetic resources. These compensations can be financial, technological, commercial, or scientific.

Different groups assign different weights to each goal, which accounts for some of the current tensions. Several northern countries as well as biologists will see in the CBD primarily a conservation convention. For others, however, namely southern countries, the second and third objectives matter most. Other groups are only interested in a fair implementation of the third goal or see in it a means of reordering relations between the state and civil society.

As in the ozone case, the CBD witnessed a new dynamics between the North and the South regarding the management of environmental issues. Now developing countries had something developed countries wanted, even though the latter would say that the issue was important for the whole world. During the negotiations, Canada sought to reconcile demands for sustainable development and the future needs of biodiversity conservation.

Developing a National Position
Canada has enjoyed an enviable reputation as a leader in environmental co-operation. For three decades, Canadians have played a pre-eminent role in promoting environ-

mental co-operation and in the development of more responsible international policies. Two of the first four executive secretaries of UNEP were Canadian. Maurice Strong, its founder, organized and chaired the two defining conferences on the environment in 1972 and 1992. James McNeill was the secretary of the Brundtland Commission whose 1987 report is largely credited for reconciling the twin goals of environmental protection and economic development through the promotion of the concept of sustainable development. Canadian delegates have stood out in many fora. Canadian scientists have contributed significantly to the promotion of ozone science and Canada has played an important role in the adoption of the Montreal Protocol. In the case of the 1984 Long-Range Trans-Boundary Air Pollution Convention negotiations, Canadians originated the 30 per cent club. The list could go on. Until recently, Canada has been perceived by the international community as an environmentally conscious country and neutral actor eager to fill the role of conciliator. This is the reputation and attitude that Canadian delegates carried into the CBD negotiations. From the beginning of the process in 1987, Canada took part in all the meetings and belonged to most expert groups. Canada was immediately open to the first UNEP invitations asking countries to make recommendations regarding the contents of a future convention.

Only 16 months separated the first meeting of the ad hoc working group of legal and technical experts in November 1990 and the conclusion of the negotiations in May 1992, which is a very short time during which to negotiate a convention of this scope. Since other treaties, such as for forests, were not proceeding as fast as was hoped originally, Mostafa Tolba insisted that the Convention be ready for the Earth Summit in Rio de Janeiro in June 1992. Thus, several issues remained unsolved at the time of signature.

If the federal government has the responsibility of negotiating treaties, biodiversity is mainly a provincial matter. Provinces will have to implement what is negotiated. Concertation, therefore, was, from the start, a basic principle that guided internal discussions on the nature of a future convention. The lead was given to the Department of the Environment and two advisory groups created in 1991: the Biodiversity Convention Advisory Group (BCAG) and the Biodiversity Convention Inter-departmental Committee (BCIC).

Government officials first met in the BCIC to discuss what should or should not appear in the CBD text. Representatives from Environment Canada, External Affairs, Indian Affairs and Northern Development, Finance, Fisheries and Oceans, and Science and Technology and from the Canadian Forestry Service and the Canadian Museum of Nature took part in the meetings. Second, the BCAG was created in order to seek advice (and reactions) from interested groups regarding the objectives the Canadian government ought to pursue and the role the Canadian delegation should play in the negotiations. The BCAG included representatives from federal and provincial governments as well as from a variety of interest groups and stakeholders. Invitations were sent to the provinces, industry (fisheries, pulp and paper, farming associations), and non-governmental organizations (Sierra Club, WWF). Most were open and responsive, with the exception of Native peoples. In 1991, while the federal government was negotiating the Charlottetown Accord, Ovide Mercredi, the Grand Chief of the Assembly of First Nations, refused to collaborate with the federal government until the status of the Native populations was revised.

One would have expected External Affairs to co-ordinate these groups since its purpose was the development of foreign policy. However, a future biodiversity convention was then perceived as a legal instrument simply to regulate the conservation of fauna and flora. It made sense, therefore, for Environment Canada to take the lead. It convened the BCAG between international meetings to discuss the state of the negotiations and prepare for the next round of negotiations. Each participant group was thus able to express its views during the course of the negotiations. In between meetings, representatives could report back to their memberships. Participating stakeholders generally viewed this process as very open and democratic. Naturally, the definition of the objectives and the political decisions made during the negotiations remained the sole responsibility of government.

The various meetings of the BCAG significantly helped Canada during the negotiations. With most relevant stakeholders around the negotiating table, the Canadian delegation was presented with a great variety of viewpoints that foreshadowed what might come up during the actual negotiations. For example, many representatives of environmental groups voiced developing countries' concerns (even though environmental and sustainable development goals do not always coincide). Industry reminded participants of the economic interests of developed countries. This process enabled Canadian negotiators to foresee potential difficulties better and prepare credible and thoughtful arguments.

As formal negotiations began, Canada was among the supporters of a CBD and eager to see the process succeed. Reasons for this enthusiasm vary. Politically, Canada was enjoying significant prestige from its role in the ozone negotiations and in the Rio process. The government also wanted international instruments that would help it protect economically important resources (such as fisheries). And as a major importer of genetic resources, Canada had an interest in securing guaranteed access to them through a formal convention. Further, officials could afford to be sympathetic to a conservation convention, believing Canada could easily comply with any of its provisions since it already had a well-developed conservation policy, particularly in terms of protected areas (Canada's principal pledge following the signature of the Convention was to expand its protected areas by 12 per cent). Indeed, Canadian agencies held one of two basic attitudes. One group wished to take advantage of the opportunity to effect change. The other believed the Convention was mainly aimed at developing countries and that Canada would not need to do anything more to live up to future obligations. Neither, then, would oppose it. The Canadian delegation as a whole was among the proactive group.

In general, Canada's position during the negotiations often leaned towards that of the developing countries. It sought to align itself squarely with the principles of conservation and sustainable development. Aware of these countries' economic stake in conserving biodiversity and of the balance of power between developed and developing countries, Canada intended to play its traditional role of facilitator and conciliator between rich and poor countries as well as that of protector of the environment. These roles are reflected in the official instructions given to the Canadian delegation in 1991, which emphasized biodiversity conservation (including fisheries), access to genetic material, technology and financial transfers to developing countries, country studies, traditional knowledge and the role of indigenous people, and national plans and strategies for biodiversity conservation.

These objectives would not evolve significantly during the negotiations. The Canadian delegation was to facilitate negotiations and seek compromises. This role was facilitated by the diversity of the Canadian delegation, which included representatives from provinces and NGOs able to work closely with specific members from other delegations.

Negotiating the CBD, 1991–1992

The negotiations proper lasted about 16 months. Each session followed roughly the same process. The IUCN would draft a proposed text, which would then be revised by various expert groups and submitted to delegations. Each article was discussed and amended until a consensus was reached. Under the general guidance of UNEP, two initial sessions of the Ad-hoc Group of Legal and Technical Experts were followed by four meetings of the Intergovernmental Negotiating Committee (INC). The first INC was almost completely stalemated by a technical point—whether it was the first or the third INC! This dispute was never resolved.

The Canadian delegation was led by Arthur Campeau, whom Prime Minister Brian Mulroney had chosen as Canada's first ambassador for the environment. The delegation included about a dozen members—six government officials, as well as representatives from the various economic sectors, NGOs, and provincial governments. Industry and NGO representatives varied from meeting to meeting to allow the largest possible participation by national groups. These non-government representatives were full members of the delegation, with the right to take part in the negotiations and speak for Canada. Indeed, Canada was the first country to include NGO representatives in its delegation and it insisted on their enjoying the same status as government representatives. Thus, NGOs felt they were more than potted plants and that they could activily help to advance various issues dear to Canada.

The Objectives of the CBD

The clash among different conceptions of the goals of the Convention emerged from the start. Should conservation or sustainable development be emphasized? Some countries, such as France and Australia, favoured developing global lists of species and areas in need of protection on which countries could base their conservation policy. Developing countries were strongly opposed to this approach, which they considered a threat to their sovereignty. As the preamble to the Convention would eventually stress, 'States have sovereign rights over their own biological resources.'

Canada was also opposed to this idea for similar reasons and preferred other incentives for biodiversity protection. Since consensus could not be reached, the idea was abandoned.

National Country Studies
An alternative was to establish independent groups of scientists who would guide countries in the process of conducting national assessments and devising national conservation strategies. Canada was indeed a pioneer in such studies. It collaborated with UNEP, the IUCN, and the World Resources Institute in creating guidelines for the conduct of country studies on biological diversity, which later were incorporated into UNEP guidelines. The Canadian Museum of Nature has helped several countries develop their monographs and strategies. Quebec has also been active in this process. This focus on national planning based on international standards characterized Canada's approach to biodiversity assessment and conservation policy. Ideally, such standards should come from an expert body, so Canada's approach was also to advocate for an international scientific advisory mechanism under the Convention, which later became the Subsidiary Body for Technical and Technological Advice.

Native Peoples
Canada also led the way in putting the concerns of Native peoples on the agenda and promoting measures that would protect traditional knowledge. More importantly perhaps, together with Sweden and Peru it introduced the concept of equitable sharing of benefits linked to the exploitation of genetic material. This was the first time such an idea was formulated in the context of an international treaty, which recognized that biological resources do not represent a mere economic potential but also have cultural significance and intrinsic value. This combination lies at the core of Native peoples' perspective, which insists on the intrinsic worth of natural resources as well as on the right to exploit them. These concerns later became embodied in article 8(j), which accentuates the political aspects of the Convention by asking parties to 'preserve and maintain knowledge, innovations and practices of indigenous and local communities . . . promote their wider application . . . and encourage the equitable sharing of the benefits arising from the utilization of such knowledge, innovations and practices.'

The notion of fair and equitable sharing of benefits raises the whole issue of property rights, potentially reorders relations between the state and society, and provides a platform for promoting the demands of Native peoples. Article 8(j) gives them the capacity to profit from their knowledge of the ecosystems they live in, and in so doing to reduce pressure on their habitat. Thus, it became an important issue after the ratification of the treaty, one whose importance is likely to grow. It is remarkable, then, that these ideas were not promoted because of intense pressures from Native populations. Indeed, since Canadian indigenous groups did not take part in the meetings of the BCAG or assign representatives to Canadian delegations, negotiators were not always on firm ground. Their instincts, however, appeared to be right. Representatives of Native peoples are now routinely included in Canadian delegations, and Canada has striven to reconcile widely differing viewpoints regarding the implementation of that article.

Although other issues (such as financial and technological transfers) occupied Canada, these were Canada's major contributions. During the course of the negotiations, the Canadian delegation kept to its early goals, concerned with protecting the interests of its Native populations and with promoting the concept of sustainable development. In that, it was not alone. Indeed, another Canadian achievement was diplomatic. Canada, Australia, and New Zealand formed a diplomatic coalition of like-minded states over some issues being discussed in the INC. As negotiations progressed, various unattached countries joined the group, in part because of the good relations that existed among some delegates. Japan and then Switzerland and the United States were included in the group. Not all states agreed on the topics under discussion, as would evidently become clear. But the existence of this group testified to the new structure of international diplomacy where negotiations take place within and among major groups of states. This group was also active in other fora and has since gained official recognition during diplomatic meetings.

Discussions proceeded at an increasingly feverish pace until the eve of the Rio Conference. The draft Convention was adopted on 22 May 1992 in Nairobi, although negotiators had left certain items to be resolved at Rio, such as the role of traditional knowledge and equitable sharing of benefits. Canadian negotiators were nevertheless pleased with the outcome, which largely followed their preference, and expected a smooth path towards formal adoption.

The United States, however, which had gone from enthusiastic promoter early in the process to reluctant player later on, announced that it would not formally sign the draft treaty. US President George Bush even asserted his intention not to go to the Rio Summit. The decision not to sign the Convention has been explained as resulting from pressures from the pharmaceutical industries concerned with intellectual property rights and unwilling to enter into any open and automatic financial arrangements around the sharing of benefits. The United States was not the only country uneasy about what it perceived as a developing-country convention more than a conservation convention. Bandwagoning on the part of other powers became a real threat. Great Britain, France, and Japan started wavering (McConnell, 1996). Members of the Canadian delegation along with others, however, thought they had conceded enough to the United States and were comfortable with the result.

On 1 June 1992, less than 24 hours after George Bush's declaration, Brian Mulroney gave a speech at the Canadian Museum of Civilization where he reaffirmed his support for the CBD and announced that Canada was ready to sign it, regardless of what the US did. Arthur Campeau had convinced Brian Mulroney that the negotiating process had broad Canadian support and that the Canadian delegation believed the Convention was sound. This had the effect of bringing fence-sitters on board. Because Canada had been a strong player during the negotiations, this position reassured many countries. Some observers saw in that move a crucial contribution to the process that prevented a reopening of the negotiations. Though 156 countries signed the Convention in Rio, the United States stood firm. The Clinton administration signed the Convention one year later but the Senate had yet to ratify it in late 2000.

After Rio

Six months later, Canada was the first industrialized country to ratify the CBD. Its own strategy was made public in 1995. In it, Canada outlined how it intended to implement the Convention, what resources it would use, and what detailed action it would undertake. The Canadian document became a kind of model that was subsequently promoted by UNEP, the IUCN, and the World Resources Institute as an example worthy of being emulated (WRI, 1995).

Montreal

The first Conference of the Parties took place in Nassau in December 1994, one year after the Convention had entered into force. One item on the agenda was the location of the permanent secretariat, for which Geneva, Madrid, and Nairobi were already candidates. The Canadian delegation took everybody by surprise (including some members of the delegation itself!) by announcing that it intended to nominate a Canadian city. There was, however, some initial confusion over which city: some members of the delegation were told to promote Montreal while the Canadian government seemed to have had Toronto in mind.

In light of these developments, it was decided to postpone a vote on the secretariat to the second COP, which was to take place a year later in Jakarta. During that time, federal and provincial delegates would each promote their preferred city. The outcome of the internal discussion appears to have had much to do with the planned Quebec referendum on independence. It seemed desirable in the eyes of Ottawa that Montreal be chosen in order to demonstrate the advantages of belonging to Canada. From then on, the federal machine went to work and organized a very efficient campaign. Within six months, the Canadian delegation managed to convince Latin American countries, which initially supported Spain, and African countries to support Montreal, which was subsequently chosen as the secretariat site. Canada again was able to demonstrate its commitment to the regime. The subsequent years, however, did not quite fulfil these early promises.

Domestic Implementation

With the election of Jean Chrétien's Liberals in 1993, Arthur Campeau was replaced by John Fraser as ambassador for the environment the following year. Most of the members of the delegation, except for the head of the Biodiversity Office, were reassigned to other positions in the civil service. Biodiversity issues gradually faded from the national and international political agenda. Eight years after signing the Convention, Parliament has yet to pass an endangered species law, as mandated by the CBD. Ottawa approved a mining project near Jasper, Alberta, close to a protected area, in apparent contravention of the Convention. Several provinces have no biodiversity strategy. Environmental activists denounced Canada's seal hunting, deforestation, animal traps, and whale hunts. In 1998, the WWF gave Canada an 'F' in regard to its implementation of the Convention and for its recent environmental record. Finally, in 1999, Canada reversed its position in the negotiations towards a biosafety protocol and took the lead of the Miami Group that stood in opposition to much of the rest of the world in defending the interest of the seed industry. In 10 years, Canada had seemingly gone from leader to villain.

Yet Canada remained active in some areas. In the biosafety negotiations, for example, Canada initially sought to play a role of mediator, trying to bridge the positions of importers and exporters of living modified organisms before allowing the interests of the biotechnology and seed industries to define its position. One must also underline Canada's continuing concern with advancing the implementation of article 8(j) on the role of indigenous populations in the face of active opposition from many countries. It supported workshops as well as the creation of an ad hoc working group on indigenous issues that reports directly to the COP. Canada also continued its action in favour of the dissemination of information by supporting the Global Biodiversity Forum, by helping developing countries participate in international discussions, and by strengthening countries' capacities to implement the Convention (for example, country monographs and strategies, and workshops on economic indicators). And, of course, it has provided substantial financial support to the secretariat. Yet, more could be done to harness Canadian expertise and use it to inform and help the regime deliver on its promises.

In Retrospect

Canada's role in the promotion of new concepts and in the negotiation of the CBD closed a remarkable decade of exceptionally active environmental foreign policy. What explains Canada's active participation in international negotiations of the CBD and the leading role it was able to play? Certainly, foreign policy objectives (coming after Canada's success in the Montreal Protocol and in climate change), domestic bureaucratic politics (the desire to internationalize certain issues so as to promote their resolution at home), a unique process of consultation that legitimized and strengthened Canada's role in the negotiating fora, solid and dedicated negotiators, and an issue that seemed initially benign to vested bureaucratic and industrial interests were all factors that may have played a role.

The dynamics of the CBD negotiations illustrate well the complexity of what Putnam (1988) has called 'two-level games' whereby negotiators must balance international and domestic interests and how each 'game' feeds and impacts on the other. Some observers have argued that these games only foster symbolic gestures. Cooper, Higgot, and Nossal (1993: 163), for example, note that despite a certain amount of leadership in international arenas on the part of Canada, there is a 'considerable degree of symbolism, driven by the domestic logic of two-level games'. The biosafety saga may illustrate one aspect of these dynamics whereby initial good intentions were thwarted at home. It and other CBD-related issues also point to the importance of involving key actors (provinces, industry, NGOs, etc.) early in the process and at both the domestic and international levels of negotiation if later diplomatic reversals and domestic implementation stalemates are to be minimized.

During the negotiation of the CBD, rather than acting as an impediment (as during the negotiations leading to the Cartagena Protocol on Biosafety), the domestic game reinforced Canada's international role. Negotiations began at home and mirrored much of what was to be expected on the international scene. If an agreement could be found at home between industry and environmental interests, between Aboriginal populations (mostly represented by government) and scientists, between development and conservation interests, then one was possible abroad. Thus, this process strengthened Canada at home (at a minimum, negotiators were able to explain their position) and made negotiators confident that they had negotiated an agreement that was acceptable to the provinces responsible for implementing much of it. And it strengthened Canada abroad, making it confident that it understood the various viewpoints and possible outcomes—and, therefore, that a compromise was possible—and enhancing its capacity to co-opt the members of its diverse delegation into rallying different constituencies.

The CBD was the first to tackle the crucial issue of who benefits and who should pay, beyond the state. This and other issues raised during the negotiating process have not disappeared; they are still being negotiated through the implementation of the accord and the discussions regarding future protocols. Canada's environmental leadership, therefore, remains crucial if only to ensure that what it successfully achieved in the course of the negotiations will actually make a difference during implementation. Canada influenced many aspects of the negotiations in a way that facilitated compromise and recognized the legitimate needs of developing countries as well as those of the indigenous populations of the world. Indeed, international environmental co-operation crucially depends on a few countries creating the conditions, investing the resources, and providing the necessary intellectual and diplomatic lead. But can one be a leader abroad and remain a laggard at home?

CONCLUSION

As this chapter has pointed out, there are domestic and international dimensions to the implementation of both CITES and the CBD. At the international level, Canada has been very active and has played a key role in fostering environmental co-operation (except in regards to the biosafety protocol). At the domestic level, its record is much poorer.

This is not as contradictory as it may seem, for both processes feed on each other. For example, domestic elements—bureaucracies, NGOs, Aboriginal communities— seek to advance their cause by 'going international', securing international commitments that Canada should honour (or that it could be accused of violating), thereby improving those groups' domestic position. The politics of animal and plant listings in CITES may reflect this process. The problem, of course, is that often those who negotiate international agreements are not those who will implement them, nor are they the ones faced with the economic consequences of trade limitations. At the same time, Canada must respond to international pressures, mostly from the European Union, that embody different (and dynamic) values about the relationship between societies and nature, especially regarding hunting and the treatment of (photogenic) animals.

Similar value-based oppositions arise in the CBD regarding the very purpose of the Convention (conservation vs sustainable development vs equity) and the role, rights, and responsibilities of indigenous groups. It is difficult for many European and African officials to comprehend the foundations and scope of the indigenous knowledge provisions of the CBD. At best, they find them irrelevant; at worst, a threat to basic political principles regarding the proper relationship between the state and society. Policy differences on the domestic and international scenes over wildlife issues thus reflect profound differences in values and political philosophies. If one adds to that potential conflicts with other international priorities, such as reducing trade barriers, one can easily despair from ever reaching solutions that would satisfy everybody's basic concerns.

Yet, Canada also enjoys advantages. Despite controversies over baby seals, leg traps, or timber, it still enjoys a positive environmental image abroad. Its scientific expertise is extensive, its presence in international organizations strong. Despite the biosafety episode it also enjoys a strong moral lead vis-à-vis developing countries that arises from its lack of imperial past, from the historical role it has played in the emergence of international environmental co-operation, from its insistence on a conservationist and sustainable development perspective that does not dissociate conservation and the welfare and rights of local groups, and from its efforts to find an international consensus over some of these issues.

The future of the international agreements on the protection of animals and plants depends on what happens domestically. In this regard, Canada has faced strong obstacles in the implementation of its international commitments. In a context where conflicting values and interests threaten paralysis, domestic pressures are often thought to be key to fostering the implementation of international commitments. That need is not without its own dangers, such as cyclical concerns, an uneven implementation of commitments, and a focus on issues and animals around which groups can be mobilized or on issues that are promoted by those groups that enjoy the largest political and financial resources. It often pits environmentalists against scientists, and environmentalists and specific economic groups against Native people. Domestic pressures, when exercised through specific clienteles, also promote bureaucratic conflicts. Finally, the nature of the Canadian federal system, which demands a certain consensus among stakeholders before action can be taken, ensures that any progress will be slow. Ultimately, the legitimacy of Canada's commitments abroad continues to be threatened by the lack of serious domestic effort in favour of habitat protection. After a first failure in 1996, the federal government introduced a new endangered species bill in April 2000 that largely relied on voluntary measures, sanctioned a transfer of property rights over endangered species from the 'public' to the private sphere, and toed a precarious

214 PART TWO: FROM SCIENCE TO POLICY

line between federal and provincial prerogatives, but this died on the order paper with the call for a November 2000 election. Both CITES and the CBD have become mainstays of Canada's conservationist image abroad, but weak internal practices and the recent influence of commercial interests raise questions about Canada's capacity to claim a moral high ground and help steer the broad biodiversity conservation effort.

NOTES

1. Peter Stoett, who attended the eleventh COP in Gigiri, Kenya, was primarily responsible for the section on CITES. He thanks the Canadian delegation to CITES, in particular Sandra Gillis, for their co-operation; Robert Campbell and several anonymous reviewers for their comments; and the Social Sciences and Humanities Research Council of Canada for funding.

2. Philippe Le Prestre was primarily responsible for the section on the CBD, which benefited extensively from interviews granted by key actors in the process and from the research assistance of Virginie Bonneau. He also thanks the SSHRC for support.

REFERENCES

Amilien, C. 1996. 'Conflicting International Parties in Tropical Timber Trade', *Environment and Conservation* 23, 1: 29–33.

Bialy, J., and B. Groombridge, eds. 1996. *1996 IUCN Red List of Threatened Animals*. Gland, Switz.: IUCN.

Boardman, Robert. 1992. 'The Multilateral Dimension: Canada in the International System', in Boardman, ed., *Canadian Environmental Policy: Ecosystems, Politics, and Process*. Toronto: Oxford University Press, 224–45.

Bryner, Gary. 1997. *From Promises to Performance: Achieving Global Environmental Goals*. New York: Norton.

Burgess, J. 1992. *The Impact of Wildlife Trade on Endangered Species*. London: London Environmental Economics Centre.

Caldwell, Lynton K. 1996. *International Environmental Policy: Emergence and Dimensions*, 3rd edn. Durham, NC: Duke University Press.

Canada. 1998. *Report of the Commissioner of the Environment and Sustainable Development to the House of Commons*. Ottawa: Minister of Public Works and Government Services.

CITES. 1999. *CITES Resolutions: Resolutions of the Conference of the Parties to CITES that remain in effect after the 10th meeting*. Lausanne, Switz.: CITES.

Colchester, Marcus. 1994. *Salvaging Nature: Indigenous Peoples, Protected Areas and Biodiversity Conservation*. Geneva: UNRISD.

Cooper, A., R. Higgot, and K.R. Nossal. 1993. *Relocating Middle Powers: Australia and Canada in a Changing World Order*. Vancouver: University of British Columbia Press.

Doern, Bruce. 1993. *Green Diplomacy: How Environmental Policy Decisions Are Made*. Toronto: C.D. Howe Institute.

Dorsey, K. 1991. 'Putting a Ceiling on Sealing: Conservation and Cooperation in the International Arena, 1909–1911', *Environmental History Review* 15, 3: 27–46.

Ehrlich, P., and A. Ehrlich. 1982. *Extinction: The Causes and Consequence of the Disappearance of Species*. London: Victor Gollancz.

Eldredge, N. 1991. *The Miner's Canary: Unravelling the Mysteries of Extinction*. New York: Prentice-Hall.

Emonsds, G. 1981. *Guidelines for National Implementation of the Convention on International Trade in Endangered Species of Wild Fauna and Flora*. Gland, Switz.: IUCN.

Environment Canada. 1999. 'Regulations Amending the Wild Animal and Plant Trade Regulations', *Canada Gazette*, Part I, 2 Oct. 1999.

Fairman, D. 1996. 'The GEF: Haunted by the Shadow of the Future', in Keohane and Levy (1996: 55–88).

Fenton, M.B. 1995. 'Species Impoverishment', in J. Leith et al., eds, *Planet Earth: Problems and Prospect*. Montreal and Kingston: McGill-Queen's University Press, 83–110.

Fitzgerald, S. 1989. *International Wildlife Trade: Whose Business Is It?* Washington: World Wildlife Fund.

Hykle, D. 1988. 'An Evaluation of Canada's Implementation and Enforcement of CITES', MES thesis, Dalhousie University.

IUCN. 1980, *World Conservation Strategy: Living Resource Conservation for Sustainable Development*. Morges, Switz.: IUCN.

Jakobeit, C. 1996. 'Nonstate Actors Leading the Way: Debt-for-Nature Swaps', in Keohane and Levy (1996: 127–65).

Johnson, A., and A. Stritch, eds. 1997. *Canadian Public Policy: Globalization and Political Parties*. Toronto: Copp Clark.

Keohane, Robert, and Marc Levy, eds. 1996. *Institutions for Environmental Aid: Pitfalls and Promises*. Cambridge, Mass.: MIT Press.

McConnell, Fiona. 1996. *The Biodiversity Convention: A Negotiating History*. London: Kluwer Law International.

McCormick, John. 1995. *The Global Environmental Movement*, 2nd edn. New York: Wiley.

McNeely, Jeffrey A. 1997. *Conservation and the Future: Trends and Options Toward the Year 2025*. Gland, Switz.: IUCN.

Mulligan, S., and P. Stoett. 2000. 'A Global Bioprospecting Regime: Partnership or Piracy?', *International Journal* (Spring): 224–46.

O'Connor, Geoffrey. 1998. *Amazon Journal: Dispatches fron a Vanishing Frontier*. New York: Plume.

Outspan Group. 1993. *The Socio-economic Value of Trade in CITES Species*. Report to the Project Scientific Advisor, CITES Administrator. Ottawa: Canadian Wildlife Service.

Putnam, Robert D. 1988. 'Diplomacy and Domestic Politics: The Logic of Two-Level Games', *International Organization* 42, 3: 427–60.

Raustiala, Kal, and David Victor. 1997. 'Biodiversity since Rio: the Future of the Convention on Biological Diversity', *Environment* 38, 4: 17–43.

Sachs, W., ed. 1993. *Global Ecology: A New Arena of Political Conflict*. London: Zed Books.

Schreurs, Miranda, and Elizabeth Economy. 1998. *The Internationalization of Environmental Protection*. Cambridge: Cambridge University Press.

Slocombe, D. 1989. 'CITES, the Wildlife Trade and Sustainable Development', *Alternatives* 16, 1: 20–9.

Stoett, Peter. 1997. 'To Trade or Not to Trade? The African Elephant and the Convention on International Trade in Endangered Species', *International Journal* 52, 4: 567–75.

Swanson, Timothy. 1999. 'Why is there a biodiversity convention? The international interest in centralized development planning', *International Affairs* 75, 2: 307–11.

Tuxill, J., and C. Bright. 1998. 'Losing Strands in the Web of Life', in L. Brown et al., *State of the World 1998*. New York: Norton, 41–58.

Weber, S. 1997. 'Institutions and Change', in Michael Doyle and John Ikenberry, eds, *New Thinking in International Relations Theory*. Boulder, Colo.: Westview Press, 229–65.

Webster, D. 1997. 'The Looting and Smuggling and Fencing and Hoarding of Impossibly Precious, Feathered and Scaly Wild Things', *New York Times Magazine*, 16 Feb., 26.

Wiese, Robert, and Michael Hutchins. 1994. *Species Survival Plans: Strategies for Wildlife Conservation*. American Zoo and Aquarium Association.

Wilson, Edward O., ed. 1988. *Biodiversity*. Washington: National Academy Press.

World Resources Institute (WRI), United Nations Environment Program (UNEP), and the World Conservation Union (IUCN). 1995. *National Biodiversity Planning: Guidelines Based on Early Experiences Around the World*. Baltimore: WRI Publications.

Young, Oran. 1989. *International Cooperation: Building Regimes for Natural Resources and the Environment*. Ithaca, NY: Cornell University Press.

The Prospects for Canada's Species at Risk

ROBERT BOARDMAN, AMELIA CLARKE, AND KAREN BEAZLEY

The task of making environmental policy in a democratic society is complex. It arises in part from what one writer has called the 'paradox of environmental policy'—'that we often understand what the best short- and long-term solutions to environmental problems are, yet the task of implementing these solutions is either left undone or is completed too late' (Smith, 1995: xiii). The problem, he suggests, is that searches for environmental solutions are unavoidably embedded in political processes. The issues of Canada's endangered species are no exception. There is abundant, though still imperfect, knowledge of the alarming status of many wildlife species. We know much about the character of the habitat changes and other threats leading to their endangerment. Crafting appropriate and effective laws, policies, and programs for the protection of species, however, has been slow. Negotiation and compromise frequently make outcomes unsatisfactory for many advocates of protection. Multiple governance systems are involved: federal, provincial, territorial, Aboriginal, municipal, and international. Actors as diverse as local wildlife groups and large mining companies bring often divergent interests and values to policy debates.

The protection of species at risk, if it is to be successful, requires attention to a diversity of concerns. It requires integration of ideas and judgements that span the biological sciences, traditional ecological knowledge, and the social sciences. Governance systems in relation to wildlife species, whether these species are used as a resource or become the focus of protection for reasons of their intrinsic value, 'have to be compatible with the character and dynamics of the ecosystems involved, and with the social, cultural, and institutional norms' of societies (Brunckhorst and Rollings, 1999: 58).

The 'norms' of society, culture, and institutions, however, do change over time; they are based on our current understanding of the nature of reality and are subject to revision as our ideas of the truth change. This is usually in response to new knowledge and to increasing tensions between how we think things ought to be and how they are. This is the situation that has resulted in the current political tensions around endangered species. On the one side, we are aware of the values of wildlife, both intrinsic and instrumental, and our desire to protect them. At the same time, we see that our actions as individuals and societies are contributing to their further declines. In this context, pressures arise for change to the attitudes, behaviours, and institutions, in other words, the norms, that are contributing to the declines and resulting in endangered species. In this way, the push for change becomes the norm and, in the future, a changed norm will evolve.

In this final chapter we look at the factors making for continuity and change over time in Canada's national system for protecting endangered species. We begin the discussion with a review of the developments that led to the federal government's plan for a Species at Risk Act (SARA) in 2000, and then discuss the main elements of the evolving national epistemic and policy communities on species at risk.

The Issue That Wouldn't Go Away: To SARA and Beyond

The collapse of the Canadian Endangered Species Protection Act (CESPA; Bill C-65) in 1997 (see Chapter 7) did not mean the end of the search for a federal legislative framework for the protection of endangered species. This became more explicitly a part of Ottawa's environmental and heritage policy after the 1997 federal election. It was already clear during the wide public debate on CESPA that that bill was only one element, though a crucial one, in the larger intergovernmental network of Canadian legal and policy arrangements on species at risk. This conception carried through to SARA. These arrangements, still in many ways in the process of being developed in the early 2000s, span the federal, provincial, and territorial governments and First Nations. They also rest on the sustained participation, or at least the tacit consent, of many organizations outside government, including private landowners, resource industries, nature trusts and other conservation groups, and scientists.

The existing policy regime continued to be developed by the federal, provincial, and territorial governments in the 1997–2000 period. At the same time, Environment Canada intensified work on the new federal bill. Provincial governments expanded their capabilities in relation to species at risk and threatened spaces, a theme we will come back to later in this chapter. The wildlife ministers and directors, representing the Canadian Wildlife Service and provincial and territorial wildlife agencies, continued to fill out and implement the two foundational agreements of 1996, the National Accord and the National Framework. Under the National Accord, ministers made a commitment to 'a national approach for the protection of species at risk. The goal is to prevent species in Canada from becoming extinct as a consequence of human activity' (Wildlife Ministers Council of Canada, 1996). The Wildlife Ministers Council of Canada agreed in October 1997 to develop a detailed plan for implementing the National Accord. The roster of provinces signing on to the accord was still incomplete. Extensive work was done in defining the terms of reference and tasks of the Canadian Endangered Species Conservation Council (CESCC), an essential step before the renovated federal bill could be introduced. CESCC is the federal-provincial body to which the Committee on the Status of Endangered Wildlife in Canada (COSEWIC), in the post-1996 regime, was to formally report its designations of species at risk.[1]

The National Accord also committed governments to develop recovery plans for endangered species (by 1997) and for threatened species (by 1998). Recovery programs have been organized since the late 1980s by the Committee on the Recovery of Nationally Endangered Wildlife (RENEW). In its tenth-anniverary report, in 1998, it noted encouraging progress in relation to several species during the decade, including the swift fox, ferruginous hawk, peregrine falcon, wood bison, and the Gaspé population of the woodland caribou. However, RENEW found that declines and serious difficulties continued to be experienced by the burrowing owl, spotted owl, Lake Erie water snake, and other species (Wylynko, 1998: 4). Also in 1997-8 RENEW reviewed the first plant recovery plan, for the wood poppy. Species recovery programs in Canada predate COSEWIC; some originated in the early work by staff of the Canadian Wildlife Service in the 1950s. Thus, recovery programs for several species, including the wood bison, whooping crane, and peregrine falcon, had been put in place before the design of a national recovery program in RENEW (Recovery Working Group, 1998, App. 2: 1; Chapter 6, this volume). Improvements to recovery processes and new approaches were the focus of RENEW meetings held in early 1998. In particular, participants called for more attention to be focused on the implementation of existing recovery plans rather than the production of new ones, and for a streamlining of procedures for developing plans (Nadeau, 1998: 4).

COSEWIC was increasingly a target of wider political controversy in the late 1990s. It nonetheless continued to act as Canada's scientific authority on species at risk. A national system for assessing the status of species and the relative risk of extinction was developed (Harper et al., 1996). Additions to the national lists by COSEWIC brought the total to 338 following the April 1999 meeting and to 353 at the May 2000 meeting in Ottawa. The committee made its first emergency listing, for the Oregon spotted frog, in November 1999.

There were conflicting opinions, however, about the place of COSEWIC in future legislation. Although it was recognized in the 1996 federal-provincial agreements as the national scientific authority for designating species at risk, key issues remained unresolved. Scientists and environmental groups expressed concern at political pressures that threatened to weaken its capacity to act as an independent voice of scientific opinion. Others, including some industry voices and provinces, preferred to tilt the balance in decisions on species listings more towards governments and to build more discretionary power into decisions on species recovery and other conservation programs, with a view to making both sets of decisions responsive not only to scientific assessments but also to economic and social factors.

The Council (CESCC) defined new terms of reference for COSEWIC in March 1999. As before, members were to include representatives from key federal agencies and from the provincial and territorial wildlife agencies. The extended structure includes eight scientific subcommittees or species specialist groups. These deal respectively with birds, terrestrial mammals, freshwater fish, marine fish, marine mammals, plants, amphibians and reptiles, and invertebrates. The chairs of these specialist groups also sit on the main COSEWIC committee. While three leading conservation organizations (the Canadian Nature Federation, the Canadian Wildlife Federation, and World Wildlife Fund Canada) had traditionally been members, the revised procedure called for CESCC to solicit nominations for these positions from the non-governmental conservation community. The revisions to COSEWIC were in effect part of a larger process of incremental change resulting from the National Accord. They also helped soothe continued provincial unease about the shape of the forth-coming federal endangered species bill. In addition, in spring 2000 COSEWIC made a decision to include an Aboriginal representative or person with traditional ecolog-ical knowledge as a member, thus accommodating one of the long-standing First Nations criticisms of the national endangered species machinery.

While this process of post-1996 regime-strengthening was under way, the new fed-eral Environment Minister, Christine Stewart, made it clear soon after the 1997 elec-tion that Ottawa intended to go ahead with its own revised legislative proposal on endangered species. Beginning in the fall of 1997 and continuing into the summer of 1999, Environment Canada officials engaged in a wide range of consultations with all of the stakeholder groups that had voiced support for or concerns about the princi-ples and detailed provisions of CESPA. The questions addressed extended to broader intergovernmental issues connected with the National Accord and Framework. The minister later described the consultations as 'among the most extensive ever done by Environment Canada' (Stewart, 1999).

Several related developments served to focus renewed attention on species at risk during the late 1990s. First, environmental organizations and scientists continued to emphasize the need for effective federal legislation (McIlroy, 1999a). These groups argued that any new bill should not duplicate CESPA. In the summer and fall of 1999, for example, the Canadian Nature Federation called on its members to write to the Environment Minister, now David Anderson, and urge two points: first, that legisla-tion should provide broader habitat protection than the restricted 'residence' criteria that had characterized CESPA; and, second, that species should be 'designated based on science rather than politics', that is, that COSEWIC rather than politicians should be responsible for making final listing decisions (*Nature Matters*, Autumn 1999). Geoff Scudder of the University of British Columbia argued strongly against giving COSEWIC only an advisory role, with final listing decisions being made by the feder-al cabinet. He pointed out that despite provincial legislation on endangered species, over two-thirds of COSEWIC-listed species were legally unprotected as a result of this kind of governmental discretion (Scudder, 2000). As they had in the mid-1990s, sci-entists again urged Ottawa to take effective action on endangered species. In February 1999, more than 600 scientists sent a letter to Prime Minister Chrétien urging strong legislation. They emphasized the vital need for habitat protection, noting that the main threat to over 80 per cent of endangered species was destruction of habitat (Johnson, 1999: 1).

Second, growing public concerns about the deteriorating condition of Canada's national parks, prompted particularly by commercial and infrastructure develop-ments affecting Banff, highlighted the urgency of tackling problems of species at risk and the need to build effective habitat-protection provisions into new legislation. The federal government's Panel on the Ecological Integrity of Canada's National Parks, chaired by Jacques Gérin, documented many of these failings in its report in December 1999. It recommended, among other things, an expanded role for scientists in questions of park management. The work of the panel led directly to the intro-duction by federal Heritage Minister Sheila Copps of a new federal parks bill in March 2000. Problems of protected areas generally and inadequate progress towards the goals set in 1992 for a national network representative of Canada's ecological regions were the focus throughout the decade of World Wildlife Fund Canada's Endangered Spaces campaign.

Third, important developments occurred in other areas of Ottawa's jurisdiction with implications for species at risk. The Oceans Act, which included sections related to marine protected areas and the protection of marine species, was passed in 1997. Developments in Canada's forestry sector converged at several points with issues aired in endangered species debates. For example, there was growing recognition in the Canadian Forestry Service (CFS) during the 1990s of the importance of integrat-ing scientific research and processes of policy formation. From the fall of 1998 the CFS expanded the traditional range of forest-related disciplines to include the promotion of linkages with the social sciences (CFS, 1999: 5). The Senate Standing Committee on Agriculture and Forestry also pressed the issue of protection of habitat following hearings in 1999 on problems of the boreal forest. Among its recommendations in June 1999 were that 'Canada needs a strong Endangered Species Act that also recog-nizes the importance of preserving the habitat on which endangered species depend for their survival, as has been the case in the United States since the 1960s' (Senate of Canada, 1999: 39). The Committee proposed that the existing powers and constitu-tional authority of the federal government in relation to fisheries, migratory birds, endangered species, and other matters be used 'to ensure a strong federal involvement in Canada's boreal forest'. The Committee also criticized the trend of national parks and other protected areas becoming isolated pockets surrounded by urbanization and development activities, and emphasized the need for these being connected, for example, through the identification and protection of key wildlife corridors. Migratory bird species fall under Ottawa's jurisdiction in the Migratory Birds Convention Act, in its current form from 1994. The federal government signed a major framework agreement with the United States in April 1997 on the development of protection programs for shared species at risk, including plans for expanded co-operation among state and provincial governments.

Fourth, the road to SARA was marked by exchanges among environmental, indus-trial, forestry, and other organizations. The work of the Species at Risk Working Group (SARWG) was central to these. Though this had no official standing, it was important in the way it defined key issues and also in the demonstration it gave of the potential for organizations with differing interests to collaborate. SARWG paralleled at the non-governmental level the formal consultations in the federal government's multi-stakeholder Endangered Species Task Force. The aim was to develop a proposal for federal legislation that would protect species and habitats while accommodating the diverse concerns of groups that had been critical of Bill C-65 in 1996–7 (May, 1999; Morbia and May, 1998). There was an early consensus that a reintroduced CESPA would be a non-starter. Representatives of several organizations took part: the Canadian Nature Federation, the Canadian Wildlife Federation, the Sierra Club of Canada, the Canadian Pulp and Paper Association, the Mining Association of Canada, and the National Agriculture Environment Committee (SARWG, 1998).

The results were summarized in a widely circulated document, *Conserving Species at Risk and Vulnerable Ecosystems: Proposals for Legislation and Programs*, in the fall of 1998. The approach was defined as preventive. That is, no species should become extinct in Canada 'because of lack of legal protection', and there should be 'no gaps in the federal/provincial/territorial safety net' (Austen et al., 1998: 2). The costs of protecting species at risk, it was argued, must be shared by all Canadians and not borne by those most closely affected, such as private landowners. SARWG envisaged two streams of protective measures. The first would focus on vulnerable ecosystems and would be a largely preventive approach to species and habitat conservation. Management plans would target multiple species in the same area. The second stream would deal with species already identified by COSEWIC as being at risk. In federal legislation, moreover, listings of species 'should be made by an independent, scientific body'. Recovery plans would involve multi-stakeholder processes and thus ensure that actions had the support of local communities. SARWG accepted that exemptions could be made, but argued against blanket exemptions of the kind that would have been possible under CESPA. Finally, the group saw a need for adequate funding for scientific research, recovery efforts, and compensation for landowners facing costs as a result of species designations and recovery programs. Requirements for tax changes were also identified, to give landowners greater incentives to support the goals of endangered species protection.

The SARWG experience thus represented a promising small-scale case study of accommodative politics. Environmental groups accepted the validity of key industry concerns, for example, on compensation for landowners, the need for socio-economic impact analysis in relation to proposed recovery programs, the prioritizing of species recovery plans, and abandonment of the idea of citizen-enforcement civil suits. Similarly, industry and resource-sector representatives moved towards positions earlier staked out by environmental groups. These included the vital requirement of protection of habitat, the science-based character of species listings, provisions for emergency listings, the need for measures such as buffer zones, and the outlawing of blanket exemption powers by the federal government. The proposals thus mirrored key aspects of the emerging approach to a national species-at-risk strategy. SARWG emphasized the National Accord and the National Framework as the foundation for species-at-risk planning, programs, and legislation. The authority of the provinces was recognized as crucial. There was also recognition of the role of First Nations in species listings and in the making and implementation of recovery plans. For example, species recovery plans, it proposed, would consider questions of sustainable levels of use for some species, the use of animal parts for ceremonial purposes, and other wildlife values held by Aboriginal cultures.

Finally, a further important context of SARA was progress among Canada's governments on agreed goals of harmonization of environmental policy. The Canada-Wide Accord on Environmental Harmonization was approved by the Canadian Council of Ministers of the Environment in November 1996, following complex negotiations begun in the spring of 1994. The agreement came very shortly after the separately negotiated National Accord and Framework and the introduction of CESPA. While the harmonization process focused on issues of environmental quality and human health and did not touch directly on wildlife and species-at-risk issues, it nonetheless underscored the important principle of an agreed division of labour among Canada's governments across all environmental policy areas. It incorporated agreement among governments that environmental tasks should be the responsibility of the level of government 'best situated' to carry them out, and recognized that a government could take environmental action within its own sphere of authority in cases where consensus on a Canada-wide approach could not be reached. Unfortunately, Aboriginal governments were not part of the negotiation process leading to the harmonization accord (House of Commons, 1997: 10).

In the 1999 Speech from the Throne, the Liberal government stated that 'Protecting species at risk is a top Government priority for the new millennium.' The Environment Minister, David Anderson, released a draft of the Species at Risk Act in December. SARA (Bill C-33) was introduced in the House of Commons in April 2000. The previous Environment Minister, Christine Stewart, had earlier said that federal legislation 'cannot ignore the established and legitimate constitutional jurisdiction of the provinces and territories'. A 'firm federal law' was 'part of the answer, but it's not the only part' (Stewart, 1999). The government thus set SARA firmly in the context of the evolving federal-provincial relationship on endangered species since the 1996 National Accord. It was described as part of a larger strategy for protecting species at risk in Canada, alongside measures by the provincial and territorial governments and encouragement to stewardship by individuals and organizations. On the last, the government stressed the need for partnerships with Aboriginal peoples and with rural communities. The 'proposed strategy sets a long-term, sustainable approach that would make incentive, stewardship and voluntary measures the preferred option to protect threatened or endangered species and their critical habitats.' Legislation would 'provide a framework for protecting species at risk as well as safety-net provisions when needed' (Environment Canada, 1999, 2000).

Many of the specific provisions of Bill C-33 represented continuity with CESPA, particularly as these reflected the principles of the 1996 National Accord. However, there were also differences. Specific provisions included prohibitions on the killing or taking of species at risk, emergency authority in relation to species or critical habitat in imminent danger, funding for species recovery and other tasks, and compensation for landowners affected by SARA decisions. COSEWIC was identified as the scientific authority, and CESCC as the body of federal and provincial ministers to which it reported (described respectively in sections 14–23 and 7 of the bill). COSEWIC was thus to be given a solid legal basis. The government stressed COSEWIC's 'arm's-length' operations and its independence, transparency, and impartiality. The procedures laid down in Bill C-33, however, called for COSEWIC only to make assessments of the status of species. These assessments would be sent to the Environment Minister and CESCC and become the basis for listing recommendations made by the Environment Minister to the federal cabinet (section 27[1–2]).

It was evident in April 2000 that Ottawa had had to walk a fine line between the positions still being defended by opponents and supporters of the specifics of the 1996 bill. Some provisions on habitat protection appeared to reverse earlier federal statements. It was politically expedient, in light of the CESPA experience, to emphasize the newness of SARA. Its title, for example, referred to 'species at risk', not 'endangered species' (which was also a more scientifically accurate description of its scope). However, to keep environmental and scientific groups on board while avoiding overt opposition from diverse private-sector interests, the federal government also had to ward off possible inflated expectations (and fears) of the bill's aims and of the powers it would give to Ottawa. Some of these concerned voices had support from ministers in the federal cabinet. Controversial provisions, such as the politicized nature of the listing process and the discretionary power of ministers in this, became in effect prerequisites for cabinet agreement on a new federal bill.

Environmental critics had mixed reactions. Some sympathized with the political constraints under which Environment Canada was operating. Most were quick to attack key elements of the draft released in December 1999 and Bill C-33 as introduced in April 2000. Stewart Elgie, of the Sierra Legal Defence Fund and a member of the federal government's Endangered Species Task Force, criticized the proposed listing process as a 'major step backwards from our current system' and from that planned in CESPA in 1996. He argued that the political and economic factors related to species protection—a major part of the government's rationale for the proposed process—should be examined in the context of specific proposals for protection and species recovery, not at the stage of listing species (Elgie, 1999: 1–2). As Sarah Dover, from the Canadian Endangered Species coalition, put it, 'Endangered species in Canada won't be able to tolerate the amount of political whimsy the minister is proposing' (McIlroy, 1999b). Other weaknesses of the bill were that, unlike CESPA, habitat protection was not 'required' even for listed species on federal lands and that Ottawa had not extended its scope to include related policy areas under federal jurisdiction, such as migratory birds and marine species. Further, although under the terms of SARA the federal government could take action where a provincial government was failing to protect a species, this again was a discretionary power; in practice, Ottawa was recognizing the provinces as having primary authority on species-at-risk matters and would be reluctant to antagonize them by intervening in this manner (Elgie, 1999: 2–5). The Canadian Nature Federation criticized the proposed legislation as the weakest endangered species bill in all of Canada's jurisdictions.

Other actors with a stake in the outcome echoed earlier concerns about federal intrusiveness into provincial matters. In light of its own endangered species legislation and protected-area planning process in 1999-2000, the government of Alberta hinted at SARA's possible unconstitutionality and raised questions about its consistency with the 1996 National Accord. Private woodlot owners in New Brunswick expressed concern about the possible implications and questioned the adequacy of and procedures for compensation (Tutton and Hrabluk, 2000). Further, by the spring of 2000 some industry and resource-sector groups were also calling for stronger species protection measures by the federal government, particularly for areas under its own jurisdiction. Many of these groups increasingly recognized the practical benefits that federal legislation could bring, for example, in reducing the chances of successful environmental group campaigns against Canadian resource products in foreign markets. The bill thus faced the possibility of either successful passage or a replay in Parliament of the multiple attacks that in effect had killed CESPA. The early call of the federal election in October 2000, however, meant that, like its predecessors, SARA, too, never ran the full legislative gauntlet.

TOWARDS A SPECIES-AT-RISK POLICY COMMUNITY

Epistemic Bases

Threats to wildlife species are a long-term feature of Canada's economy. New factors include the effects on species abundance, diversity, and distributions of anthropogenic climate change. Knowledge of these factors and effects has grown rapidly in Canada over the last decade. Significant data gaps remain in many areas, however, for example, in relation to invertebrate species and, as Martin Willison noted in Chapter 5, regarding marine as opposed to terrestrial species. Methodological issues remain on crucial issues such as the classification of Canada's landscapes, a complex task increasingly associated with consideration of the activities and concerns of resource users (Wiken, 1998: 2–3).

However, science-based conservation efforts have been vulnerable to the criticism that the definitions and criteria of knowledge in such appraisals are themselves limited. Two very different instances can be noted. First, some critics who object to species-at-risk protection on grounds of its economic costs take issue with what is argued to be a misleading use of terms like 'species' and 'endangered' in conservation debates. COSEWIC lists, for example, include subspecies and geographical populations as well as species; and while COSEWIC distinguishes between 'endangered' and 'extirpated' species, wide use of the former term is held by critics to carry with it the erroneous implication that any species that disappears from Canada has become extinct, even if it thrives in the US part of its range (Jones, 1999: 7–8).

Second, Aboriginal critics have questioned the adequacy of conventional scientific approaches to wildlife (Barsh, 2000: 155). While the federal government's national strategy included the commitment that Ottawa would 'continue to work closely with Aboriginal peoples to ensure their participation in species assessment and recovery efforts' (Environment Canada, 1999), no specific mechanisms were proposed for the integration of traditional knowledge in the procedures of COSEWIC or in the development of recovery plans. To achieve this objective, the Nunavut Wildlife Management Board had earlier proposed that two members with traditional knowledge should be added to COSEWIC. One criticism of SARA was that it excluded Aboriginal representatives from key decision-making bodies. Ottawa was prepared to assimilate in legislation 'the components of our knowledge they deem useful, all the while ignoring the world-views, values, philosophies and intellect that reside within the holders of our traditional knowledge.' SARA thus 'marginalizes Aboriginal peoples and their knowledge' (Simpson, 2000: 8–9). The long-delayed decision to include in COSEWIC in 2000–1 a member with traditional Aboriginal ecological knowledge went some way towards meeting these criticisms, but the underlying complaint persisted of an imbalance in the relative contributions of traditional and conventional scientific insights and values in species listings and recovery.

Questions of knowledge—of the adequacy of data and the methods by which it is arrived at—have thus been inseparable from species-at-risk practices and processes. Sustained pressures from environmental and scientific groups characterize the endangered species epistemic community. Federal and provincial governments might, hypothetically, have taken legislative and other actions if this political context were missing or attenuated. Governments, after all, have their own scientists and wildlife agencies. Without continual prodding and reminders from environmental and scientific groups, however, it is doubtful that even the existing components of the national regime would have materialized.

Although organizationally distinct, environmental and scientific groups tend to have broadly similar views on species at risk and to promote comparable conservation objectives based on ecological principles. The perspective is shared by other groups, such as the Canadian Bar Association, which has its own environmental law section. It has 'long been of the view that, within the scope of federal jurisdiction, Parliament should adopt legislation providing effective protection for endangered species and their habitats' (Burgess and Proudfoot, 1996: 2). Environmental groups are significant, too, as conservation actors with their own programs on species recovery and the protection of habitat. For example, in 1999 WWF Canada introduced 55 recovery plans for species and populations listed as nationally endangered or threatened.

In relation to the federal government's legislative planning, the main aim of these groups has been to raise public awareness of the need for habitat protection and sound ecosystem management. Federal legislation, as WWF Canada stated in relation to CESPA, has to 'acknowledge simple ecological principles' (WWF, 1996). Restricted definitions of habitat, such as those focusing on the nest or den of a species, are of limited value from an ecological perspective. Such an approach is 'like protecting your bedroom, but allowing the rest of your house and neighbourhood to be bulldozed' (Smith, 1999). On these criteria, environmental critics argued that in SARA the federal government had failed to deliver on its commitments to protect species at risk and their habitats, even on lands clearly within federal jurisdiction (Smallwood, 1999).

Competing Process Models

Effective habitat protection based on sound principles of ecosystem management, then, is a requirement of species conservation. The implications in terms of protected-area planning for such issues as ecological connectivity and wildlife corridors have been much debated in scientific and environmental bodies. The fundamental requirement of habitat protection, however, transforms the protection of species at risk into a much more politicized issue. In public perceptions, protecting a rare flower or a photogenic wild bird may be a desirable objective, even one defended with emotion, but it can remain somewhat abstract. Protecting the habitat it needs to survive, however, immediately raises practical questions of costs and choices. This might entail changes in land use or zoning, increased government regulation, a risk of job losses resulting from a mandated curtailment of economic activity, and declines in property values. From this perspective, the scientific determination of the status of species and analyses of the factors threatening them constitute only one aspect of an overall policy process. Consideration of the economic costs and social and political significance of proposed measures to protect species is, for many critics, crucial to this larger process. An extension of this position, vigorously opposed by conservation groups and the scientific consensus, is that such considerations should enter the listing process itself.

A large number of stakeholders thus continue to have interests in Canada's endangered species regime. Many Aboriginal, environmental, industrial, forestry, and agricultural actors, as well as the provincial and territorial governments, responded—largely negatively—to Ottawa's plans for endangered species legislation in 1996–7. Labour unions have followed the issue closely and voiced concerns about the economic implications of species protection measures. On some issues in relation to CESPA, labour groups supported environmentalists' positions; on others, including the question of compensation, their views were closer to those of representatives of industry and private landowners (Harris, 1997; White, 1997). Kim Pollock, of the Industrial Wood and Allied Workers of Canada, wrote in connection with SARA planning that 'environmental initiatives carry an economic and social price tag. When land is removed from production, whether to maintain habitat for endangered species or for other reasons, economic costs are likely to arise.' Pollock cites US industry estimates of 29,000 forest jobs lost as a result of protection of the northern spotted owl. And 'since it is often well-heeled, urban-based environmental non-governmental organizations that advocate species initiatives, those groups should declare right off the bat how much they will pay to implement plans. Too often in the past, environmentalists' appetites have been large, mainly because they did not expect to pay for the meal' (Pollock, 1999: 2).

Landowners, from households and private woodlot owners to large companies, have also argued that proposals from environmental groups for protection of species at risk often ignore inequities in the burden of costs. The issue has divided environmentalists. Some question why it should be necessary to pay someone to protect the natural environment. Exchanges among groups in the late 1990s, for example, in the SARWG framework, led to increased appreciation of this problem on the part of environmental organizations and to support for the principle that the costs of species protection should be shared by all Canadians. Species recovery, as one Alberta official has put it, is 'a broad societal goal that can't always be effectively achieved simply by wildlife agencies'. It has to include the support and participation of landowners (Steven Brechtel, quoted in Wylynko, 1997: 7). Estimates made in the conservation community in the late 1990s, based in part on exchanges among environmental groups and forestry associations, were that around $150 million per year would be required to fund species protection; the February 2000 federal budget announced $90 million over three years for various aspects of species protection.

While an important focus of debate on species at risk has moved to consideration of the positive incentives to landowners that can ensure their long-term support for conservation goals, the core issue of compensation has persisted. The importance of encouraging stewardship by private landowners had been acknowledged in earlier federal-provincial policy statements, including particularly the national wildlife policy of 1990. CESPA was nonetheless criticized because it 'had little to say about the flavour of agreements with private landowners and managers' and did not confront the issue of compensation (Forsyth, 1997: 6). The Canadian Federation of Woodlot Owners has argued that the tax system and forestry regulations act as disincentives to woodlot stewardship. The size of lots, it argued in testimony to the Senate Agriculture and Forestry Committee in 1999, is such that adequate income is not generated, particularly in view of the time trees take to grow and the fact that many operations are part-time (Senate of Canada, 1999: 34). The New Brunswick Federation of Woodlot Owners indicated in April 2000 in initial reactions to SARA that the issue of compensation remained crucial (Tutton and Hrabluk, 2000).[2]

A further concern voiced in relation to the CESPA framework was that citizen enforcement actions could harm landowner interests and also set back recovery programs. According to the Alberta Forest Products Association (1997), species would not be protected if the government was spending time and money on court cases. Vexatious or frivolous civil actions could be brought forward, resulting in the legal harassment of private landowners regardless of whether the law had been broken (Canadian Cattlemen's Association, 1996).

Landownership and management questions have also been central to the concerns of nature trusts and conservancy organizations. These have developed a variety of options to promote positive incentives for stewardship, including the negotiation of easements with private landowners and the ownership or management of lands with significant ecological values. These programs are crucial to national species conservation goals of promoting representativeness in provincial protected-area systems, building wildlife corridors and other ecological connections among important habitats, and encouraging landowner participation in species recovery projects. The approach has fostered debate on tax laws, for example, capital gains tax relief on donations of land to nature trusts or following conservation easement agreements. Ottawa's ecological gifts program has reduced the capital gains tax for donations of ecologically sensitive land. Comparisons with the more conservation-friendly US tax laws reinforce pressures for reforms in Canada. US laws have assisted nature trusts to acquire or manage critical habitats. The US Nature Conservancy, for example, has acquired land and easements totalling 1.4 million hectares and valued at about $2.7 billion (Mittelstaedt, 2000).

Jurisidictions and Species-at-Risk Governance

Designing the process of listing species at risk raises fundamental issues of the relations between science and policy. Scientific and environmental groups have consistently argued that the procedure for listing a species as threatened, vulnerable, endangered, or extirpated should be exclusively a scientific task and should not be mixed up with questions of the economic and social consequences of recovery options. Awareness of the implications of species listing decisions, however, gives governments a powerful incentive to take the matter into their own hands or at least to retain a decisive say in listings changes. Of the provinces that had endangered species legislation or provisions in place by 2000, six give provincial cabinets final listing authority. In the case of Nova Scotia's legislation in 1998, however, any species listed by COSEWIC that is native to the province has to be included in the provincial species-at-risk list (Keith, 1999: 4). The former approach, however, became the model for the SARA listing process. Their role in listing decisions is only one of the ways in which provincial governments exercise critical authority in relation to the overall framework for protecting Canada's species at risk.

Jurisdictional questions continue to be central to national approaches to endangered species. SARA was designed as part of a national framework, grounded in the principles of the National Accord and National Framework of 1996, that recognizes the provinces as key players. Environmental groups argued throughout the 1990s for a strong and effective federal role in the national safety net for endangered species. This is essential to avoid piecemeal approaches and a patchwork of measures across the country that may contain significant gaps (Canadian Endangered Species Campaign, 1998: 1). The provinces vary in their legal and political commitments to habitat protection and species recovery, some laws have serious defects, and, as noted above, for the most part provincial governments, as opposed to their scientific advisory committees, retain power over species designations. In a globalizing economy, moreover, provinces compete with each other and may be tempted to weaken environmental regulations to attract outside investment and promote exports.

For its part, Ottawa has critical incentives to be demonstrably effective in protecting species at risk. These arise from its role as Canada's national government and the public expectations associated with this, the process of playing catch-up with other Western countries whose national governments already have endangered species legislation in place, and its need to be seen in the international community to be implementing commitments such as those entered into in the Convention on Biological Diversity in 1992. Part of the federal government's aim has also been to amplify the existing framework of federal legislation that touches on species and protected-area questions. This includes the Canada Wildlife Act, the Migratory Birds Convention Act, the Oceans Act, and the Wild Animal and Plant Protection and Regulation of International and Interprovincial Trade Act.

However, expectations on the part of conservation groups that the two federal legislative initiatives, CESPA and SARA, would, if successful, presage a stronger leadership role for Ottawa in the Canadian governance system were unfulfilled. First, each was based on recognition of the roles of other governments, the definitions of tasks set out in principle in the 1996 National Accord, and the pivotal role of the CESCC. Second, there have been differences of view and counter-pressures inside the federal government. Key departments have promoted the interests of their respective constituencies. Agriculture officials, for example, have respected the concerns about endangered species legislation of the Canadian Federation of Agriculture and the Canadian Cattlemen's Association. Third, the direct authority of the federal government on species-at-risk issues is restricted to lands under its jurisdiction. Responding to the inadequacies of SARA on this last point, Leanne Simpson writes that she imagined a protected owl and its habitat: 'I sat there quietly imagining the piece of land where the owl lived—with the owl, its nest, the tree and a large clear-cut surrounding the tree like an ocean. But then I laughed . . . realizing that the piece of land I had been imagining could only be on an "Indian Reserve" or in a national park' (Simpson, 2000: 8–9).

Developments in the 1990s continued to underline the central role of provincial governments. These continued to build their respective endangered species and protected spaces frameworks. Nova Scotia passed its endangered species legislation in 1998. Quebec elaborated its strategy on protected areas in 1998, though it remained critical of key aspects of the National Accord. It also continued to list species, including 10 plant species designated as threatened or vulnerable (*Eco*, 1998: 12). Alberta undertook extensive consultations in 1999–2000 on its planned system of protected areas. Ontario's 'Lands for Life' round tables from early 1997 focused unprecedented public and media attention on habitat protection needs in the province and eventually resulted in recommendations for a significant expansion and consolidation of parks and protected areas. Other measures filled out specific aspects. For example, the 1998 publication by the Ontario Forest Research Institute of a comprehensive plant list for the province was a significant aid in the future protection of plant species at risk (Obenchain, 1998: 8).

These kinds of initiatives reinforce the image of the provinces as the foundational actors on endangered species and protected areas. This perspective was evident in provincial opposition to Ottawa's CESPA initative in 1996–7. Governments maintained this would have led to unacceptable intrusions into provincial jurisdiction, significantly altered the nature of the provinces' partnership with Ottawa, eroded their longer-term capacities to manage provincial economies, and undermined the ethic of the National Accord (Amos, 1999: 113–16). The position was supported by many private-sector groups. The Mining Association of Canada (1996) argued that Ottawa 'should be as effective as possible within its recognized scope of authority', but that the use of 'powerful tools . . . to supersede the authority of the provinces should be discouraged.' The National Accord is thus the basis for national species-at-risk policy development as far as the provincial governments are concerned (Hagan, 1998). Specifically, through their membership in CESCC, provincial ministers retain important powers in relation to the composition of COSEWIC and in deliberations on its species assessments.

A final element of continuity and incremental development is provided by activities at the international level. Creation of a federal law for endangered species was a commitment following Canada's signature of the CBD (see Chapter 9). The resulting Canadian Biodiversity Strategy was nonetheless criticized by conservation groups on the grounds that it did not 'commit governments to develop legislation, contain[ed] no time-lines and place[d] the protection of endangered species in jeopardy' (Sierra Club of Canada, 1999: 3). The design of implementing legislation, however, has proved much more difficult and protracted than for Canada's adherence to CITES. The participation of Canadian scientists in international programs with conservation aspects, for example, in the IUCN/World Conservation Union, is similarly an important context of Canada's evolving national framework on species at risk. Further, Canada's northern research activities, which too often have a low or unfocused priority for governments, have significant species components. These are situated in the larger international framework of circumpolar scientific and conservation activities. It is often forgotten, too, that Canada also participates in bipolar science through Antarctic research as a non-consultative party to the Antarctic treaty; the Protocol on Environmental Protection that entered into force in 1998 includes strict rules for wildlife protection in the region (Roots, 1998: 6).

CONCLUSION

Attitudes towards wildlife have historically occupied an important place in Canadian society and culture. Wildlife species have been used in Canadian fiction as an 'extended metaphor for a lost past, escape from the present and a return to a state of being which breaks through linguistic, cultural and psychological boundaries' (Harger-Grinling and Chadwick, 1998: 53). Residues of the traditional image of species abundance have persisted into the present, and this image is often resistant to arguments made by environmentalists and scientists that habitat protection is urgently needed, especially in the urbanized and developed southern regions of Canada. As Larry Simpson of the Nature Conservancy of Canada has put it, 'The perception of most Canadians is that we have a big country and there are more resources over the next hill, so what's the panic?' (Mittelstaedt, 2000).

The evidence of widespread public support for federal endangered species legislation suggests this comforting myth may at last be losing its appeal. 'People from all walks of life, including farmers, environmentalists, miners, public servants, foresters, fishers, bankers, and bakers recognize the importance of the conservation and sustainable use of the environment and specifically the protection of endangered species' (Forsyth, 1997: 6). Whether this support is arising out of an intrinsic or instrumental valuing of endangered species is unclear; however, either origin will suffice for the present to justify their protection. That said, a large shift is still required to link our desires—as well as our ethical responsibilities to endangered species and future generations—to concrete policy and other action.

What is required, and may be beginning, is widespread public debate on the issue and subsequent changes in societal values and institutions. The conservation of endangered and other species, whether for their own sake, for our enlightened self-interest, or for future generations, is, in the end, the most efficient, equitable, and cost-effective way for humanity to live sustainably into the future. To protect the species and habitats that exist is, ultimately, the most 'conservative' of actions, though at this point in our social development it may also seem to be the most radical.

Although the scientific knowledge around endangered species remains imperfect, there is more than enough evidence that the crisis is indeed real and potentially catastrophic and that the problem is primarily caused by human activities. For example, there is general agreement that large areas of habitat need to be protected, far more than is currently found in parks and other protected areas. The obstacles to a meaningful response to the science and the crisis are political. Philosophical and scientific arguments are both clear in demonstrating the need to protect endangered and other species. At a broad societal level, the value of endangered and other species is recognized. Yet, at a political level, the conflict remains in questions of how to implement this ideal. These issues and others will continue to play a prominent political role. We are still far from resolution of the conflicts and solutions. Multiple views persist about key issues of jurisdiction, financing and compensation, the role of science in listing and recovery activities, and the appropriate balance between regulatory and voluntary mechanisms for pursuing conservation goals. The effectiveness of a sustained commitment for the future depends on how well the diverse stakeholders can work together. The future of Canada's species at risk lies in the balance.

NOTES

1. COSEWIC at first (from the late 1970s) reported to the annual federal and provincial/ territorial wildlife conference, and from the late 1980s to the wildlife ministers and directors. It was a central feature of the National Accord and Framework of 1996 but remains without a legal basis until such time as federal endangered species legislation can be passed.

2. Bill C-33 stated that the minister may pay compensation 'to any person for losses suffered as a result of any extraordinary impact' of the listing of a species or an emergency order on habitat protection (section 64[1]).

REFERENCES

Alberta Forest Products Association. 1997. Submission to the House of Commons Standing Committee on Environment and Sustainable Development.

Amos, W.A. 1999. 'Federal Endangered Species Legislation in Canada: Explaining the Lack of a Policy Outcome', MA thesis, University of British Columbia.

Austen, Catherine, et al. 1998. 'A "Made in Canada" approach', *Recovery: An Endangered Species Newsletter* (Fall): 2.

Barsh, Russel Lawrence. 2000. 'Taking Indigenous Science Seriously', in Stephen Bocking, ed., *Biodiversity in Canada: Ecology, Ideas, and Action*. Peterborough, Ont.: Broadview Press, 153–74.

Brunckhorst, David J., and Nick M. Rollings. 1999. 'Linking ecological and social functions of landscapes: I. Influencing resource governance', *Natural Areas Journal* 19, 1: 57–64.

Burgess, J.M., and Gordon Proudfoot. 1996. Letter to Hon. Sergio Marchi and Hon. Allan Rock. Ottawa: Canadian Bar Association.

Canadian Cattlemen's Association. 1996. Proposed Revisions to Bill C-65. Calgary: CCA.

Canadian Endangered Species Campaign. 1998. *It's Time to Act! 1998 Report Card on the Implementation of the National Accord for the Protection of Species at Risk*. Ottawa: CESC.

Canadian Forestry Service. 1999. *Solutions: The CFS Newsletter* (Spring-Summer).

Eco. 1998. Issue #12 (Nov.): 12.

Elgie, Stewart. 1999. *Analysis of Proposed Canadian Species at Risk Act*. Toronto: Sierra Legal Defence Fund.

Environment Canada. 1999. *Canada's Plan for Protecting Species at Risk*. Ottawa: Environment Canada.

———. 2000. *The Species at Risk Act: A Guide*. Ottawa: Environment Canada.

Forsyth, Sheila. 1997. 'The challenge of protecting species', *Recovery: An Endangered Species Newsletter* (Spring): 6.

Hagan, Doug. 1998. *A Report on the Progress of Canadian Jurisdictions in Protecting Species at Risk: The Response to the National Accord*. Peterborough, Ont.: Ontario Ministry of Natural Resources.

Harger-Grinling, Virginia, and Tony Chadwick. 1998. 'The wild animal as metaphor in the narrative of three examples of Canadian fiction', *Etudes canadiennes* 44: 53–60.

Harper, B., et al. 1996. 'Ranking', *Recovery: An Endangered Species Newsletter* (Fall): 6.

Harris, Deborah Gudgeon. 1997. 'Federal Act halted in Parliament', *Recovery: An Endangered Species Newsletter* (Spring): 7.

House of Commons. 1997. *Harmonization and Environmental Protection: An Analysis of the Harmonization Initiative of the Canadian Council of Ministers of the Environment*. Report of the Standing Committee on Environment and Sustainable Development. Ottawa: House of Commons, Dec.

Johnson, Marc. 1999. 'Scientists demand government action on endangered species habitat protection', *Nature Alert* 9, 2 (Spring): 1.

Jones, Laura, with Liv Fredricksen. 1999. *Crying Wolf? Public Policy on Endangered Species in Canada*. Vancouver: Fraser Institute.

Keith, Todd. 1999. 'The Wilderness Act and the Endangered Species Act become law', *Natural Landscapes* 5, 1: 3–4.

May, Elizabeth. 1999. Personal communication with Amelia Clarke, 28 Oct.

McIlroy, Anne. 1999a. 'Tough endangered species law demanded', *Globe and Mail*, 24 Feb., A1, A11.

———. 1999b. 'Ottawa's endangered species plan called inadequate', *Globe and Mail*, 18 Dec., A3.

Mining Association of Canada. 1996. Submission to the House of Commons Standing Committee on Environment and Sustainable Development, 21 Nov.

Mittelstaedt, Martin. 2000. 'Conservationists push changes to tax laws', *Globe and Mail*, 24 Jan., A7.

Morbia, Rita, and Elizabeth May. 1998. 'Unlikely allies join to protect Canada's species at risk', *Global Biodiversity* 8, 3 (Winter): 19–21.

Nadeau, Simon. 1998. 'Improving the national species recovery process', *Recovery: An Endangered Species Newsletter* (Spring): 4.

Obenchain, Abigail M. 1998. 'OFRI publishes comprehensive list of Ontario's plant species', *Insights* (OFRI) 3, 2 (Summer): 7–8.

Pollock, Kim. 1999. 'Species protection in the balance', *Recovery: An Endangered Species Newsletter* 13 (June): 2.

Recovery Working Group. 1998. Recovery Planning and Implementation for Species at Risk: Improving the System in Canada (draft). Ottawa: Environment Canada.

Roots, E.F. 1998. 'The Environmental Protocol now in force: What does it mean?', *Newsletter for the Canadian Antarctic Research Network* 6 (May): 6.

SARWG. 1998. *Conserving Species at Risk and Vulnerable Ecosystems: Proposals for Legislation and Programs*. Ottawa: Species at Risk Working Group.

Scudder, Geoff. 2000. 'Our Wildlife are on Life Support', *Globe and Mail*, 21 Feb., A15.

Senate of Canada. 1999. *Competing Realities: The Boreal Forest at Risk*. Standing Senate Committee on Agriculture and Forestry, Subcommittee on the Boreal Forest, Issue No. 19 Pt. I. 56th Parliament, 1st Session. Ottawa: Senate of Canada.

Sierra Club of Canada. 1999. *Endangered Species: Going, Going, Gone . . . Why Canada Needs Endangered Species Legislation*. Ottawa: Sierra Club of Canada.

Simpson, Leanne. 2000. 'First Nations and last species', *Alternatives* 26, 1 (Winter): 8–9.

Smallwood, K. 1999. E-mail, 17 Dec. Vancouver: British Columbia Endangered Species Coalition.

Smith, J. 1999. <www.cnf.ca/species_sci.html> Ottawa: Canadian Nature Federation.

Smith, Zachary A. 1995. *The Environmental Policy Paradox*, 2nd edn. Englewood Cliffs, NJ: Prentice-Hall.

Stewart, Christine. 1999. 'Working together on species at risk', *Recovery: An Endangered Species Newsletter* 13 (June).

Tutton, Michael, and Lisa Hrabluk. 2000. 'Bill doesn't go far enough, say environmentalists', Saint John *Telegraph Journal*, 12 Apr., A8.

White, R. 1997. Submission by the Canadian Labour Congress to the House of Commons Standing Committee on Environment and Sustainable Development.

Wiken, Ed. 1998. 'Classifying Canada's landscapes and seascapes', *Eco* 12 (Nov.): 2–3.

Wildlife Ministers Council of Canada. 1996. *National Accord for the Protection of Species at Risk*. Ottawa: Canadian Wildlife Service.

World Wildlife Fund. 1996. Submission to the House of Commons Standing Committee on Environment and Sustainable Development.

Wylynko, David. 1997. 'The individual's role: Endangered wildlife a priority for all', *Recovery: An Endangered Species Newsletter* (Winter): 7.

——. 1998. 'Celebrating a 10-year anniversary', *Recovery: An Endangered Species Newsletter* (Spring): 4.

Web Sites for Endangered Species/ Species-at-Risk and Wildlife Habitat

(Note: Many of the Web sites listed were provided courtesy of Wildlife Habitat Canada and the Canadian Council on Ecological Areas.)

Agriculture and Agri-Food Canada
   http://www.agr.ca
Alberta Forest Products Association
   http://www.abforestprod.org
Alberta Government
   http://www.gov.ab.ca
   http://www.gov.ab.ca/env/parks/anhic
   http://www.glimr.cciw.ca/wildspace/ims-intromap,html
Alberta Registered Professional Foresters Association
   http://www.arpfa.org
Alberta Wilderness Association
   http://AlbertaWilderness.ab.ca/
Alberta's Threatened Wildlife
   http://www.gov.ab.ca/env/fw/threatsp/index.html
Arctic Institute of North America
   http://www.ucalgary.ca/aina/research/research.html
Association of British Columbia Professional Foresters
   http://www.rpf-bc.org
Atlantic Coastal Zone Information Steering Committee
   http://is.dal.ca/aczisc/aczisc
Atlantic Region State of the Environment Reports
   http://www.ns.ec.gc.ca/reports/soe.html
Biodiversity in Canada (S. Bocking)
   http://www.trentu.ca/biodiversity
British Columbia Endangered Species Coalition
   http://www.wcel.org/esc/

British Columbia Government
  http://www.gov.bc.ca
  http://www.elp.gov.bc.ca/sppl/soerpt/trends
British Columbia Government—Forests
  http://www.gov.bc.ca/bcgov/key/nat.htm
British Columbia Ministry of Environment, Land and Parks Wildlife Branch
  http://www.elp.gov.bc.ca/wld/
British Columbia Protected Areas
  http://www.env.gov.bc.ca/sppl/soerpt/protect/pa3.html
Canada Mortgage and Housing Corporation
  http://www.cmhc-schl.gc.ca/cgi
  n/webc.exe/mktinfo/store/National_Market_Analysis.html#chs
Canadian Arctic Resources Committee
  http://www.carc.ca
Canadian Biodiversity Information Network
  http://www.cbin.ec.gc.ca/Biodiversity
  http://www.cbin.ec.gc.ca/Biodiversity/en/Search/SidebarSearch.cfm
Canadian Centre for Inland Waters
  http://www.cciw.ca
Canadian Council of Forest Ministers
  http://www.ccfm.org/home_e.html
Canadian Council of Ministers of the Environment
  http://www.mbnet.mb.ca/ccme/
Canadian Council on Ecological Areas
  http://www.cprc.uregina.ca/ccea
  http://cprc.uregina.ca/ccea/ecozones/index.html
Canadian Endangered Species Coalition
  http://www.chebucto.ns.ca/Environment/FNSN/hp-cesc.html
Canadian Federation of Agriculture
  http://www.cfa-fca.ca/enviro-e.htm
Canadian Forestry Service
  http://nrcan.gc.ca/cfs/proj/ppiab/sof/common/latest.shtml
Canadian Institute of Forestry
  http://www.cif-ifc.org
Canadian International Development Agency
  http://www.acdi-cida.gc.ca/index-e.htm
Canadian Museum of Nature
  http://www.nature.ca
  http://www.nature.ca/english/atrisk.htm
Canadian Nature Federation
  http://www.cnf.ca/species_sci.html
Canadian Parks and Wilderness Society
  http://www.cpaws.org/

Canadian Plains Research Centre
   http://www.cprc.uregina.ca
Canadian Polar Bear Commission (Canadian Polar Bear Information Network)
   http://www.polarcom.gc.ca/cpin.htm
Canadian Pollution Prevention Information Clearinghouse
   http://www.ec.gc.ca/cppic
Canadian Pulp and Paper Association
   http://www.open.doors.cppa.ca/english/facts/forets/index-ok.htm
Canadian Soil Information System
   http://res.agr.ca/cansis/_overview.html
Canadian Wildlife Federation
   http://cwf-fcf.org/
Canadian Wildlife Service
   http://www.ec.gc.ca/cws-scf/cwshom_e.html
Canadian Wildlife Service—Marine Protected Areas
   http://www.cws.scf.ec.gc.ca/habitat/marine/Eindex.html
Circumpolar Arctic Flora and Fauna
   http://www.grida.no/prog/polar/caff/welcome.htm
Commission for Environmental Co-operation
   http://www.cec.org/programs_projects/conserv_biodiv/index.cfm?varlan=english
Committee on the Status of Endangered Wildlife in Canada
   http://www.cosewic.gc.ca/COSEWIC/Default.cfm
Compendium of Canadian Forestry Statistics
   http://nfdp.ccfm.org/frames2_e.htm
Convention on Wetlands of International Importance
   http://www.ramsar.org
Ducks Unlimited
   http://www.ducks.ca
Ecological Monitoring and Assessment Network
   http://www.cciw.ca/eman/intro.html
EcoMap
   http://atlas.gc.ca
   http://www-atlas.ccrs.nrcan.gc.ca/projects/atlas/v2/default.htm
Environment Canada
   http://www.ec.gc.ca
   http://www.ec.gc.ca/envhome.html
Environment Canada, Atlantic Region—Sensitivity Mapping Program
   http://www.ns.ec.gc.ca/mapping/index.htm
Environment Canada—Clean Water
   http://www.ec.gc.ca/envpriorities/cleanwater_e.htm
Environment Canada—Climate Change
   http://www.ec.gc.ca/climate/ccs/pdfs/volume7.pdf
   http://www.ec.gc.ca/climate/ccs/ccs_e.htm

Environment Canada, Yukon Region
  http://www.pyr.ec.gc.ca/index_e15ab.htm
Environment Canada Ecosystem Initiatives
  http://www2.ec.gc.ca/ecosyst
Environment Canada Indicator Series
  http://199.212.18.79/~ind/English/TOC/toc_e.htm
Environment Canada Infobase
  http://www1.ncr.ec.gc.ca/~soer/ANEF/default_e.html
Environmental Index
  http://www.stud.ntnu.no/~hall/green.html
Federation of Canadian Municipalities
  http://www.fcm.ca
  http://www.fcm.ca/_vti_bin/shtml.dll/index.html
Federation of Ontario Naturalists
  http://www.ontarionature.org
Fisheries and Oceans Canada
  http://www.ncr.dfo.ca
Fisheries and Oceans Canada—Freshwater Institute
  http://www.cisti.nrc.ca/programs
Fisheries Resource Conservation Council
  http://www.ncr.dfo.ca/frcc/fisheries/2/index.html
Foreign Affairs and International Trade
  http://www.dfait-maeci.gc.ca
  http://www.dfait-maeci.gc.ca/english/foreignp/environ/menu.htm
Forest Engineering Research Institute of Canada
  http://www.feric.ca
Fraser River Action Plan (Environment Canada, Pacific and Yukon Region)
  http://www.pyr.ec.gc.ca
  http://www.pyr.ec.gc.ca/ec/frap
Georgia Basin Ecosystem Initiative
  http://www.pyr.ec.gc.ca/GeorgiaBasin/gbi_eIndex.htm
Great Lakes 2000
  http://www.cciw.ca/glimr/program-e.html
  http://www.cciw.ca/solec/intro.html
Great Lakes Information Management Resource
  http://www.cciw.ca/glimr/state-e.html
Indian and Northern Affairs Canada
  http://www.inac.gc.ca
  http://www.inac.gc.ca/natres/index.html
Industrial, Wood and Allied Workers of Canada
  http://www.iwa.ca
Industry Canada
  http://www.ic.gc.ca

Institute for Marine Biosciences
   http://www.nrc.ca/cgi-bin/corporate/external.pl?http://www.nrc.ca/imb/
Institute for Pacific Ocean Science and Technology
   http://www.ipost.org
International Arctic Environment Data Directory
   http://www.dpc.dk/ADDGreenlandDK/DKmaps.html
International Development Research Centre
   http://www.idrc.ca
International Institute for Sustainable Development
   http://iisd1.iisd.ca/nr.htm
International Union for the Conservation of Nature and Natural Resources/
World Conservation Union
   http://www.iucn.org
Land Owner Resource Centre
   www.lrconline.com
Man and the Biosphere Canada
   http://www.cciw.ca/mab/intro.html
   http://www.cciw.ca/cbra/english/page6.html
Manitoba Conservation Wildlife Branch
   http://www.gov.mb.ca/natres/wildlife/
Manitoba Government
   http://www.susdev.gov.mb.ca
Manitoba Government—Dept of Water Resources
   http://www.gov.mb.ca/natres/watres/wrb_main.html
Manitoba Government—Fisheries
   http://www.gov.mb.ca/natres/fish/fish.html
Manitoba Government—Forestry
   http://www.gov.mb.ca/natres/forestry/index.html
Manitoba Government Land Information Division
   http://www.gov.mb.ca/natres/lid/index.html
Manitoba State of the Environment Report
   http://www.gov.mb.ca/environ/pages/soe91/legisl.html
Maritime Lumber Bureau
   http://www.mlb.ca
Meewasin Valley Authority
   http://www.lights.com/meewasin
Millennium Eco-communities
   http://www.ec.gc.ca/eco/main_e.html
Mining Association of Canada
   http://www.mining.ca
   http://www.mining.ca/english/initiatives/index.html
Ministère de l'Environnement et de la Faune du Québec
   http://www.mef.gouv.qc.ca/en/index.htm
   http://www.fapaq.gouv.qc.ca/

National Accord for Protection of Species at Risk
   http://www.cws-scf.ec.gc.ca/sara/strategy/accord_e.htm
National Atlas
   http://www-nais.ccm.emr.ca/
National Atlas of Canada
   http://www.atlas.gc.ca
National Forest Strategy Coalition Secretariat
   http://www.nfsc.forest.ca
National Round Table on the Environment and the Economy
   http://www.nrtee-trnee.ca
   http://www.nrtee-trnee.ca/eng/programs/aboriginal/aboriginal-bulletin2_e.htm
   http://www.nrtee-trnee.ca/eng/programs/health/health_e.htm
   http://www.nrtee-trnee.ca/eng/programs/sustainable_cities/sustainablecities_e.htm
National Water Research Institute
   http://www.cciw.ca/nwri-e/intro.html
Natural Resources Canada
   http://www.nrcan.gc.ca
   http://www.nrcan.gc.ca/homepage/nat_resources.shtml
   http://www.nrcan.gc.ca/homepage/nat_resources.shtml.geoconnexions.org/
      english/index.html
Natural Resources Canada—Canadian Forest Service
   http://www.nrcan.gc.ca
Natural Resources Canada—Fire Report
   http://www.nrcan.gc.ca/cfs/proj/sci-tech/arena/firereport_e.html
Nature Conservancy of Canada
   http://www.natureconservancy.ca
New Brunswick Government
   http://www.gov.nb.ca
Newfoundland Government
   http://www.gov.nf.ca
North American Waterfowl Management Plan
   http://www.wetlands.ca/nawcc/nawmp
Nova Scotia Department of Natural Resources, Wildlife Division
   http://www.gov.ns.ca/natr/wildlife/
Nova Scotia Government
   http://www.gov.ns.ca
Nunavut Environmental Database
   http://136.159.147.171/scripts/minisa.dll/20/1/0?SEARCH
Nunavut Government
   http://npc.nunavut.ca/eng/npc
Ocean Voice International
   http://www.ovi.ca/status.html
Ontario Government
   http://www.gov.on.ca

Ontario Ministry of Natural Resources
  http://www.mnr.gov.on.ca/MNR/index.html
Ontario Professional Foresters Association
  http://www.opfa.on.ca
Ontario's Niagara Escarpment
  http://www.escarpment.org
Ontario's Species at Risk
  http://www.rom.on.ca/ontario/risk.html
Ordre des ingénieurs forestiers du Québec
  http://www.oifq.com
Parks Canada
  http://www.parkscanada.pch.gc.ca/parks
Prairie Farm Rehabilitation Administration
  http://aceis.agr.ca/pfra/pfintroe.htm
Prince Edward Island Government
  http://www.gov.pe.ca
Quebec Forest Industries Association
  http://www.aifq.qc.ca
Quebec Government
  http://www.gouv.qc.ca
Quebec Lumber Manufacturers Association
  http://www.sciage-lumber.qc.ca
Quidi Vidi Foundation
  http://199.212.18.76/ecoaction/success/story6_e.htm
Ramsar Convention on Wetlands
  http://www.cws-scf.ec.gc.ca/habitat/ramsar/eindex.html
Recovery of Nationally Endangered Wildlife
  http://www.cws-scf.ec.gc.ca/es/renew/index_e.html
Saskatchewan Government
  http://www.gov.sk.ca
Saskatchewan Wetland Conservation Corporation
  http://www.wetland.sk.ca
Sierra Club of Canada
  http://www.sierraclub.ca/
Sierra Legal Defence Fund
  http://www.sierralegal.org/
Species at Risk
  http://www.cws-scf.ec.gc.ca/sara/main.htm
Species at Risk Act
  http://www.parl.gc.ca/36/2/parlbus/chambus/house/bills/government/
    C-33/C-33_1/C-33_cover-E.html
St Lawrence Vision 2000 Action Plan
  http://www.slv2000.qc.ec.gc.ca

Standing Committee on Environment and Sustainable Development
http://www.parl.gc...t-1996-11-12_a.html
State of Canada's Environment Infobase
http://www1.ec.gc.ca/~soer
http://www3.ec.gc.ca/~ind/English/Home/default1.htm
State of Canada's Forests
http://nrcan.gc.ca/cfs/proj/ppiab/sof/common/latest.shtml
State of Parks
http://parkscanada.pch.gc.ca/library/DownloadDocuments/documents_e.htm
State of the Environment Reports for Manitoba
http://www.gov.mb.ca/environ/pages/soerepts.html
Statistics Canada
http://www.statcan.ca/english/Pgdb/Land/geogra.htm#lan
Survey on the Importance of Nature to Canadians
http://www.ec.gc.ca/nature
The Nature Conservancy
http://www.heritage.tnc.org
Toronto and Region Conservation Authority
http://www.trca.on.ca
United Nations Environment Program
http://www.unep.org
United States Fish and Wildlife Service
http://www.fws.gov
United States Fish and Wildlife Service—Border States
  Alaska Fish and Wildlife Service—Region 7
    http://www.r7.fws.gov/
  Idaho Department of Fish and Game
    http://www.state.id.us/fishgame/fishgame.html
  Maine Inland Fisheries and Wildlife
    http://janus.state.me.us/ifw/homepage.htm
  New York State Department of Environmental Conservation
    http://www.dec.state.ny.us/website/dfwmr/index.html
  US Fish and Wildlife Service—Region 3
    (Minnesota-Wisconsin-Illinois-Indiana-Michigan-Ohio)
      http://www.fws.gov/r3pao/
  US Fish and Wildlife Service North—Region 6
    (North Dakota-Montana)
      http://www.r6.fws.gov/
  Vermont State Fish and Wildlife
    http://www.anr.state.vt.us/fw/fwhome/index.htm
  Washington Department of Fish and Wildlife
    http://www.wa.gov/wdfw/

United States—State of the National Ecosystems
    http://www.us-ecosystems.org
University of Guelph
    http://www.crle.ouguelph.ca/
University of Regina
    www.cprc.uregina.ca/ccea/ecozones
Urban Habitat Stewardship Resource Network
    http://www.gatewest.net/~cwhp/
Wild Species at Risk in Saskatchewan
    http://www.serm.gov.sk.ca/ecosystem/speciesatrisk.php3
Wildlife Habitat Canada
    http://www.whc.org
Wildlife Habitat Canada—Report on the State of Wildlife Habitats in Canada
    http://www.whc.org/hsr/report-e.htm
Wildlife Management Advisory Council (North Slope)
    http://www.taiga.net/wmac
Windstar Wildlife Institute
    www.windstar.org/wildlife
World Wildlife Fund Canada
    http://www.wwfcanada.org/
Yukon State of the Environment Report
    http://www.taiga.net/yukonsoe/

Contributors

WILLIAM AMOS is a special assistant to the federal Environment Minister, the Honourable David Anderson. He advises the minister on the Species at Risk Act and water issues, as well as on western regional environmental issues. He completed his MA in Political Science at the University of British Columbia in August 1999.

KAREN BEAZLEY is an assistant professor in the School for Resource and Environmental Studies at Dalhousie University. Her research and teaching interests include protected-area system planning, biological diversity conservation, conservation biology, landscape ecology, and environmental ethics. She is a member of the board of directors of the Science and the Management of Protected Areas Association (SAMPAA).

ROBERT BOARDMAN is a professor of Political Science at Dalhousie University. His research interests focus on international environmental policy. His publications include *Canadian Environmental Policy: Context and Cases for a New Century* (2001), co-edited with Debora VanNijnatten.

STEPHEN BOCKING is an associate professor in the Environmental and Resource Studies Program at Trent University. His research interests include environmental history and the role of science in environmental politics. His books include *Ecologists and Environmental Politics: A History of Contemporary Ecology* (1997), and an edited collection: *Biodiversity in Canada: Ecology, Ideas, and Action* (2000).

AMELIA CLARKE is a Masters of Environmental Studies candidate at Dalhousie University. Her research interests include environmental management systems for universities and forest ecosystem conservation. She is the founder and former National Co-ordinator of the Sierra Youth Coalition, an executive committee member of the Sierra Club Atlantic Canada Chapter, and on the steering committee of the Canadian Environmental Network—Forest Caucus.

PHILIP DEARDEN is a professor of Geography at the University of Victoria where he teaches courses in environmental management, protected areas, and

Southeast Asia. He has been an adviser on protected area issues to the Asian Development Bank, World Bank, IUCN, and various governments throughout the world and has ongoing research programs in Southeast Asia and Canada on marine parks, ecotourism, and ecosystem-based management. He is co-author of *Parks and Protected Areas in Canada: Planning and Management* (1993) and *Environmental Change and Challenge: A Canadian Perspective* (1998).

PETER EWINS directs WWF Canada's Species and Arctic Programs. He has worked extensively with governments to implement species recovery plans and to streamline the recovery planning system in Canada. His doctoral work was on seabird ecology in the Shetland Islands and he worked for six years for the Canadian Wildlife Service on Great Lakes wildlife toxicology programs.

BILL FREEDMAN is a professor and chair of the Biology Department at Dalhousie University. His research interests involve the ecological effects of pollution and disturbance, with a recent focus on the effects of forestry activities. Recent publications include *Environmental Science: A Canadian Perspective*, 2nd edn (2000), and *Environmental Ecology*, 2nd edn (1995). He is a member of the board of directors of the Nature Conservancy of Canada and of the Tree Canada Foundation.

DAVID GAUTHIER is a professor of Geography and executive director of the Canadian Plains Research Centre, University of Regina. His current research and teaching areas include rural social cohesion, protected ecological areas planning and management, applications of GIS in natural resources management, ecological land classification, and human adaptations to climate change. He is treasurer of the Canadian Council on Ecological Areas, a trustee of the Nature Conservancy of Canada, and a member of the World Commission on Protected Areas, the National Atlas of Canada steering committee, the Grasslands National Park Management planning committee, the Saskatchewan Prairie Conservation Action Plan steering committee, and numerous professional organizations. He also directs the University of Regina Press and serves as an editorial board member for the *Great Plains Journal*.

DAVID M. GREEN is an associate professor and the curator of vertebrates at the Redpath Museum of McGill University. His research interests include the evolutionary genetics, ecology, and behaviour of amphibians and the principles of population biology relating to population declines. He is chair of the Committee on the Status of Endangered Wildlife in Canada (COSEWIC), president-elect of the Society for the Study of Amphibians and Reptiles (SSAR), and on the board of directors of the IUCN's Declining Amphibian Populations Task Force.

KATHRYN HARRISON is an associate professor of Political Science at the University of British Columbia. She is the author of *Passing the Buck: Federalism and Canadian Environmental Policy* (1996) and co-author, with George Hoberg, of *Risk, Science, and Politics: Regulating Toxic Substances in Canada and the United States* (1994).

GEORGE HOBERG is associate professor of Political Science and Forest Policy at the University of British Columbia. He is author of *Pluralism by Design: Environmental Policy and the American Regulatory State* (1992); co-author, with Kathryn Harrison, of *Risk, Science, and Politics: Regulating Toxic Substances in Canada and the United States* (1994); co-editor of *Degrees of Freedom: Canada and the United States in a Changing Global Context* (1997); co-author of *In Search of Sustainability: Forest Policy in British Columbia in the 1990s* (2001); and editor of *North American Integration: Economic, Cultural, and Political Dimensions* (2001).

PHILIPPE LE PRESTRE is professor of Political Science at the University of Quebec at Montreal (UQAM), where he specializes in international environmental politics and foreign policy analysis and heads the Global Ecopolitics Observatory. Between 1995 and 1999 he chaired the Environmental Studies Section of the International Studies Association. His publications include *Écopolitique Internationale* (1997).

LINDSAY RODGER is a species recovery manager with World Wildlife Fund Canada. She is WWF's representative to the Committee on the Status of Endangered Wildlife in Canada and works on the development and use of effective tools for species-at-risk recovery, including recovery planning and implementation, legislation, and community stewardship. She authored *Tallgrass Communities of Southern Ontario: A Recovery Plan* (1998) and is the founding chair of the Ontario Tallgrass Prairie and Savanna Association.

PETER J. STOETT is assistant professor of International Relations at Concordia University, where he specializes in global ecopolitics and human rights issues. Recent publications include *Human and Global Security* (2000) and *The International Politics of Whaling* (1997).

DAVID T. SUZUKI is a professor in the Sustainable Development Research Institute at the University of British Columbia. After a career as a geneticist, he has worked on ecological issues and with Aboriginal peoples in many parts of the world. He is a recipient of the Order of Canada and numerous other awards. With his wife, Dr Tara Cullis, he co-founded the David Suzuki Foundation, a science-based organization that seeks solutions to root causes of ecological degradation and communicates those solutions to a wide audience.

ED WIKEN is the managing director of the Habitat Status Program at Wildlife Habitat Canada and the chair of the Canadian Council on Ecological Areas. Ecosystem management and applications in fields such as state of the environment reporting, indicators, and protected areas constitute some of his professional interests. He has authored/co-authored over 200 scientific publications, many of which remain as benchmark ecosystem-level studies of audits, sustainability, management, monitoring, and mapping. He has worked extensively throughout Canada and in other countries, including Zimbabwe, Russia, Mexico, and the US.

MARTIN WILLISON is a professor of Biology and Environmental Studies at Dalhousie University, where he teaches nature conservation and protected areas management. In his research and advocacy conservation work he focuses on the marine environment and associates with several non-government organizations, some of which he helped to found.

Index

Blanding's turtle, 89, 91, 99
Botanical gardens, arboreta, 91
Bottlenose whale, 105–6
Bouchard, Lucien, 132
Bowhead whale, 82, 101, 102
Brundtland Commission: *See* World Commission on Environment and Development
Burrowing owl, 82, 83, 219
Butterflies, 27, 42, 170, 171

Caccia, Charles, 144
Campeau, Arthur, 207, 209
Canada-Wide Accord on Environmental Harmonization, 155, 223
Canada Wildlife Act, 124, 139, 230
Canadian Bar Association, 226
Canadian Biodiversity Strategy, 131, 190, 231
Canadian Council of Ministers of the Environment, 154–5
Canadian ecozones, 32, 55, 58
Canadian Endangered Species Coalition, 148, 161, 224
Canadian Endangered Species Conservation Council, 218, 219–20, 223–4, 230, 231
Canadian Endangered Species Protection Act: *See* Bill C-65, (CESPA)
Canadian Environmental Assessment Act, 153–4
Canadian Environmental Law Association, 132
Canadian Environmental Protection Act, 153, 159
Canadian Forestry Service, 221
Canadian Nature Federation, 27, 127, 132, 222, 224
Canadian Ocean Habitat Protection Society, 110
Canadian Parks and Wilderness Society, 88, 125, 131–2
Canadian Pulp and Paper Association, 222
Canadian role internationally, 4, 192–3, 197–9, 201, 205, 212–14; *see also under* Convention on Biological Diversity
Canadian Wildlife Federation, 27, 126, 220, 222
Canadian Wildlife Service, 3, 57, 128–9, 132, 140–1, 195–6; recovery plans, 29, 121, 128–9, 219; research and education, 122–4, 129
Captive breeding programs, 121, 123
Caribou, 22, 23, 77, 82, 121, 129, 191, 219
Carson, Rachel, 123
CBD: *See* Convention on Biological Diversity
CESCC: *See* Canadian Endangered Species Conservation Council
CESPA: *See* Bill C-65, (CESPA)
Chretien government, 148, 151, 210, 223
Circumpolar Agreement on the Conservation of Polar Bears, 125, 191

CITES, 2, 4, 37–8, 125, 127, 133, 139, 181, 182, 190–2, 193–200, 231; Canadian role, 197, 198, 199; Conference of Parties, 193, 196–7, 199–200; lists, 193, 197; *see also* Wild Animal and Plant Protection and Regulation of International and Interprovincial Trade Act
Citizen suits, 140, 141, 143, 151, 155, 160
Climate change, 1, 133, 171, 225
Committee on the Status of Endangered Wildlife in Canada: *See* COSEWIC
Conservation easements, 37, 183, 228; *see also* Tax relief conservation incentives
Conservation of Terrestrial Ecosystems program, 129
Conservation on private land: *See* Stewardship conservation
Conservation values/value shift, 75, 120, 123, 124, 133–4, 217–18, 231–2; *see also* Biodiversity values
Convention for the Preservation of Wild Animals, Birds, and Fish in Africa, 201
Convention on Biological Diversity, 2, 4, 131–3, 137, 140, 181, 190–1, 192, 200–14, 229, 231; biodiversity country studies, 208; Canadian position development, 205–7; Canadian role, 192–3, 200–1, 204, 206–11; ecosystem protection approach, 202, 204; indigenous peoples issues, 207, 208–9, 211; Montreal Secretariat, 210; NGO participation, 207; US role, 192, 209–10
Convention on International Trade in Endangered Species of Wild Fauna and Flora: *See* CITES
Convention on Wetlands of International Importance, Especially as Waterfowl Habitat: *See* Ramsar Convention
Convention Relative to the Preservation of Fauna and Flora in Their Natural State, 193
Copps, Sheila, 86, 141, 147, 221
Coral reefs, 108–9; *see also* Deep-sea corals
COSEWIC 27, 119, 126–8, 133, 219–20, 231; Aboriginal participation, 27, 220, 226; committees, 27, 219; designation process, 28, 218; NGO representation, 27, 220; politics, 107–8, 141, 148, 219, 227, 229; reports process, 27
COSEWIC status list, 26–32, 39–48, 95, 128; changes, 29; legal standing, 30, 36, 139, 141–2, 148, 161, 218, 220, 224; taxonomic groups, 29–31; under-represented species/taxa, 26
Cost compensation: *See under* C-33, (SARA); Bill C-65, (CESPA)
Crown lands, 154, 155, 156
Cumulative effects, 2, 56, 107
CWS: *See* Canadian Wildlife Service

Species at Risk Working Group, 158, 221–3
Species decline, 2, 12
Species value: aesthetic, 16; God's creation, 17; intrinsic, 18; life-support, 15–16; Noah principle, 18; resources, 16–17
Species *vs* individual value, 18
Spruce budworm spraying, 122
Stewardship conservation, 27, 54, 83, 130–1; incentives, 36–7, 38, 83, 158–9, 160–1, 222, 223, 228–9; by NGOs, 131, 161, 172–3, 175, 184–5, 228–9; OECD, 183–5; *see also* Nature trusts; NGO conservation programs
Stewart, Christine, 158, 223
Strong, Maurice, 205
Sustainable development approach, 203–4
Suzuki, David, 147
Switzerland, 173, 179, 182
Swordfish fishery, 105
Symposium on Canadian Endangered Species and Habitats (1976), 127, 132

Task Force on Endangered Species Conservation, 141
Tax relief conservation incentives, 37–8, 83, 222, 228; Ontario Conservation Land Tax Incentive Program, 83
Threatened species, 3, 28
Tolba, Mustafa, 203, 205
Tourism, 120–1, 125
Toxic pollutants, 21, 26, 29, 35, 56, 129, 132, 170; long-distance effects, 52–3, 87, 104, 122
Traditional knowledge, vii, 217, 222, 225–6
Transboundary conservation programs, 182–3
Trumpeter swan, 121

Umbrella species, 23, 88
United Kingdom, 170, 179
United Nations Conference on Environment and Development, 131, 140, 203, 205, 209–10
United Nations Environment Program (UNEP), 191, 197, 203, 205, 208, 210
United States, 171, 173–4, 176, 180–1, 182, 183–4; Endangered Species Act, 95, 127, 137, 140, 150, 157, 174, 175–6, 180–1; Fish and Wildlife Service, 29, 30, 34, 176, 177
Urbanization, 3, 170, 221

Vascular plants, 79; *see also* Plants

Vertebrate endangered species, 26, 31–2, 170
Vulnerable species, 3, 12, 20, 22, 29

Wetlands, 36, 122, 125, 170
Whales and whaling, 33, 100–2, 104, 125, 192; *see also* individual species
White pelican, 78
Whooping crane, 77, 80–1, 121, 219
Wild Animal and Plant Protection and Regulation of International and Interprovincial Trade Act, 139, 230
Wildlife and wilderness as resources, 121, 177
Wildlife corridors, 88, 221
Wildlife enjoyment, 120–1
Wildlife Habitat Canada, 38, 51, 130
Wildlife management and protection: federal, 52, 131; provinces, 30, 56, 82–3, 120–1, 124–5; provincial jurisdiction authority, 4, 139–40, 154, 155, 156
Wildlife Ministers Council of Canada, 141, 153, 218
Wildlife protection, provinces, 30, 56, 82–3, 120–1, 124–5
Wildlife research, 121, 122–4, 129
Wildlife trade, 193, 197–8; Canadian, 195–6, 198; trade regulation, 199
Wolves, 79, 81, 85, 87
Wood buffalo, 52, 77, 128, 219
Wood Buffalo National Park, 77, 79–81, 86, 121
Woodland Caribou Provincial Park, 82
World Charter for Nature, 203
World Commission on Environment and Development, 131, 203
World Conservation Strategy, 203
World Conservation Union: *See* IUCN
World Resources Institute, 208
World Trade Organization, 198
World Wildlife Fund, 172, 199, 202, 203
World Wildlife Fund Canada, 3, 27, 38, 89, 94–5, 105, 125, 127, 128, 131–2, 161, 205, 211, 220; recovery programs, 29, 36, 89
World Wildlife Fund for Nature: *See* World Wildlife Fund

Yellowstone to Yukon (Y to Y) campaign, 88
Yoho National Park, 86

Zoos, 191, 199